Before the Beginning, During the Middle, After the End

Before the Beginning, During the Middle, After the End

Cosmology, Art, and Other Stories

Lucian Krukowski

PICKWICK Publications · Eugene, Oregon

BEFORE THE BEGINNING, DURING THE MIDDLE, AFTER THE END
Cosmology, Art, and Other Stories

Copyright © 2014 Lucian Krukowski All rights reserved Except for brief quotations in critical publications or reviews, no part of this book may be reproduced in any manner without prior written permission from the publisher Write: Permissions, Wipf and Stock Publishers, 199 W 8th Ave., Suite 3, Eugene, OR 97401.

Pickwick Publications
An Imprint of Wipf and Stock Publishers
199 W 8th Ave., Suite 3
Eugene, OR 97401

www.wipfandstock.com

ISBN 13: 978-1-62564-599-9

Cataloging-in-Publication data:

Krukowski, Lucian, 1929–

    Before the beginning, during the middle, after the end : cosmology, art, and other stories / Lucian Krukowski.

    p ; cm

    ISBN 13: 978-1-62564-599-9

    1. Cosmology—Philosophy. 2. Metaphysics. 3. Aesthetics. I. Title.

BD541 .K78 2014

Manufactured in the U.S.A.   08/13/2014

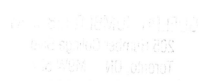

For Samantha, Zoë, and Ute

Contents

- I First Things | 1
- II Three Themes | 4
- III Divisions | 9
- IV Cosmologies | 17
- V Passages | 24
- VI Birds | 30
- VII Ontological Questions | 36
- VIII Images as Portents | 44
- IX Antecedents | 51
- X Preamble to Making Art | 60
- XI Proposals for Making Art | 62
- XII Perplexities in Making Art | 67
- XIII Ways of Making Art | 71
- XIV Reasons for Making Art | 80
- XV A Return to Origins | 89
- XVI Savanna | 101
- XVII Edvard and Umma | 110
- XVIII Walter | 121
- XIX Anna | 172
- XX Pain and Pleasure | 202
- XXI Beginning, Middle, End | 228

I
First Things

THE WORLD CONTINUES ON while we attest to its coming and going by writing and singing, by making art, war, and love. How our world began and how it will end are less clear than how it is being lived. The first and last of these are imprisoned in theory—conjectures about how (if) it was before beginning began—and inferences to the state of affairs after we, or something like us, ends. Neither state of affairs—if they are such—has a witness. Of course, conjecture and inference look for support—whether this comes from the mystical, rational, or empirical camps of thought. But the stretch back is awfully long, and the notion of nothing into something is obdurately vague. We do not even know whether anything ever did begin; we only know that about ourselves. Equally, when we talk about the end of the world, we still seem only to be talking about ourselves—for who (or what) else could describe what remains after the world (our world) has ended. We can extend the difficulty by saying that we have no knowledge about what preceeded the beginning of the universe (which includes our world as a proper part), nor do we have any idea of what the universe is like after all cognition (of our kind) has disappeared.

To find some clarity—and also retain our uniqueness—we wrap the world around ourselves by speaking of creation, redemption, and judgment, and of a source, a god or a process, as cause and reason for these transformations. In this sense, we and the world are alike in origin and conclusion in that something, of which we know little, provides for the mysteries of the beginning and the end.

I write here about three themes, which have to do with the passage of our world. I first identify them in the theological sense—as Creation,

2 Before the Beginning, During the Middle, After the End

Annunciation, and Judgment. These are religous terms which, in one form or another, have been central in coping with the mysteries of being—the worlds we recall, inhabit, and anticipate. I do not endorse the cogency of these terms through my discussion, but I do offer them as a good vantage for a historical path into the limits of existence, and as support for my later attempts to find imagistic equivalents in both the history—and practice—of art.

Questions about the beginning, middle, and end of the world are typically approached through philosophic conjecture or scientific analysis. But such themes also have currency in art—not as questions evidently, but as evocative images. Here, I take a critical and hermeneutic approach within which themes of process and purpose can be located in artworks through their stylistic histories and ambitions. To further this, I indicate how present art, when open enough to reconsider or reconstitute such themes, could change the nature of today's efforts to give art polemical purposes, and so provide other (better?) reasons for its making. I conclude this writing with some stories, unevenly biographical, partly fictional, which can be taken as parables for the developed themes and their transformations. The aim here is to elucidate a view of art as providing symbols of specific sorts for beginning, living, and ending.

This is a large order. But we who are agnostic, and so have less youth but more patience than do others, need to embrace large panoramas against which to pit our tools of invention and insight—even as we (grudgingly) try to stay within the limits of what we know. Notwithstanding these and other limits, I will project my best images of this journey into otherness through the transformation of my three themes into questions of how they came to be, what their internal changes are, and where their futures lie. However imaged, I try to keep these themes free of acrimonies, whether in the market place or in formal polemics, that would constrict their lives. Rather, as with all sources of good images—these themes should hover over but not rest on their material origins—nor be limited to the beliefs they have variously exemplified. I believe that themes and artworks should both be suspicious of chronology—they should not seek their justification in the uncertain histories of their makers or their times. These are best rethought with every reading or looking.

But images, like persons, have a fear of freedom: Does an image (or a theory) want to free itself from the security, and the limitations, of its material sources—and wander into what, for it, is not thinkable? Do persons (even those we esteem for their derring-do) ever free themselves from the supportive, yet restrictive, matter of the beliefs that they select to form their lives? Imagine Leibnitz's disappointment at reading Carnap—or

Michaelangelo's at seeing a Rothko. But, as well, can such persons acknowledge—perhaps to their greater advantage—the impetus behind these beliefs, and so form a picture of the antinomic power these beliefs need to succeed themselves?

Even with these caveats—which I accept but do not submit to, for I have often said (to unbelievers) that I do not really care about the weighty lessons of the past or the looming trivialities of the future. But I write this in my own late time; and so must try—as do those others who are primed to bypass the inertia of old bodies and stagnant styles—to write as if free of place and precedent: We, you and I, will, as we must—no?—engage the grandiosiest themes to be found in our imagination, and then suffuse the unwarned world with their juices. Juices are mostly good! We cannot worry that interpretations of these themes do not conform to extant theories. "Extant" is good enough for those who are not habituated to parse and pick at how the world is—who do not admit to a warranted dislike of "present" (when was that?) affairs. I try, here, to say more than I have said before—to suggest new options about how it all first came together; and to suggest a standpoint, or two, from which to treat the question of what it all is like after it ends.

Themes, to be public, do not always need material bodies—but they can often be seen or heard, and sometimes smelled, felt and tasted. Then there are the other, more speculative, themes which do not want to be so identified—they are at two as regards the senses. We have a long history of ideas without bodies—which, when we think of it, has often caused the most grievous hurts to the bodies that they disdain. But here, although I shudder at the slaughters perpetrated in the name of disembodied principles, my aim is not recrimination. I want to put ideas and bodies into a continuum that is aware of past histories—including the histories that attempt to extend their own interests through special pleading, past the later times that offer contrary evidence. The subjects of my themes not only derive from physical history, they are also about speculative realities that extend beyond the before and after of our world. But these are realities we would want to, but cannot yet, maintain through causal claims that encompass the farthest parameters of what we call existence.

II

Three Themes

THE DIVISIONS THAT MARK my subject are three: The first is that point where the world begins—where it appears from out of the mystery of non-being. The second lies somewhere between its progeny and its future—that point between beginnings and ends where we, the beneficiaries of our being-here come together to sing a celebration of the wonder that it happened at all, and then intone the fear of its ending. The third division is a speculation on ways that mark the end—our own and the ending of the world. I use these to consider the logical difficulties of locating a something that comes from nothing (an uncaused cause), onto a linear historical tradition that imposes a value on the progression (progress?) of that something, and so requires a judgment which imparts a (final) verdict onto all that has happened before. I then look at putative distinctions between "before, now, after" to see if concepts of stasis, duration, and change can be made broad enough to clarify what it means to each to say the world began, is enduring, and will end. I first offer these through categories that are consonant with religious attempts to invest life and history with purpose—for these form the major explanatory traditions of western culture, and are a thematic source of much of its greatest art.

"Beginning" goes halfway from nothing to something. Haydn began his "Creation" with a slow murmuring that wasn't going anywhere until a glorious major chord showed it the way to go. "Middle" is a somewhere perched between both sides, but it is also the moving, yet persistent, boundary between worlds that are antithetical: the kiss Caravaggio has Judas plant on the cheek of Christ. "End" is not a finality—only the unremarked moment when light fades and (our) sentience stops. It is an image brought beyond uttermost simplicity. If an image at all, it is one of unremitting dark—of silence or a black canvas—and is accompanied by the question of what all that was all about.

Creation

The first theme—Creation—is about transition between the before and after of the world's beginning. Michaelangelo depicts God separating the light from the dark, and creating out of this, the sun, the earth, and the heavens. This theme carries the metaphysical state of being into the physical realm of what happens when "being begins"—although Michaelangelo does not show the condition before the advent of creation. Some theories hold that this transition is seamless and ongoing, and that the dividing prefixes "non" and "pre" are atavisms of a compartmentalized thinking—giving inordinate weight to the temporality of human consciousness. Other theories, perhaps professing the rigor of dialectic, expect the recognition of (perhaps discontinuous) totalities to remain a mode of explanation—even when lacking an enabling agent for the continuance between them. Theories of this kind point to "self-consciousness" as the evolutionary goal of developmental process. Hegel calls this process "Evolution of the Spirit"—a final (post-temporal) transcending of material limitation, and the end—and completeness—of the dialectic process. Jean Paul Sartre, more agnostically, uses such Hegelian terms as "an sich" (en-soi), "für-sich" (pour-soi), to show the expansion of consciousness from lower to higher biological forms—and into human self-consciousness. Sartre is not concerned with the physics, but rather with the the transition between consciousness to self-consciousness to social consciousness (pour-nous). The dialectic Sartre posits invariably generates an "other" to anything that is determined to be the case. (a guiding rationale for revolution). So even if there is no "non" or" pre" to beginnings, there is something (perhaps the something of nothing) that provides a site for its intersection with actuality—a something which offers a promise of both discord and progress.

The advent of creation is more mysterious than its end, for it involves the "ex nihilo" aspect that confounds, but also lubricates, our speculations about causality. But it is difficult to pair nothing and something through a causal link—unless we find other concepts that help us bypass the difficulty of considering nothing as also a causal agent. Such concepts would have to hold nothing to be (or have) a different kind of something—thus requiring a new expression in both physical and metaphysical theory which affords a linkage (or obviates the distinction) between nothing and something—in our case, between pre-being and being.

We do not know the circumstances of the Big Bang; we were not there, and can only speculate about the nature of this state—if indeed there was one. But what preceded it—an infinite series of states of non-being? To avoid this threat of corrosive regress may require a freeing of thought from the limits of linearity, and a willingness to follow ideas that refute the very notion of a time-directed origin in favor of other notions—like that of a steady state of infinite change (a fashionable oxymoron), or Nietzsche's cyclical "eternal return."

The creation of the world, so another story goes, became an affront to the unitary but previously complacent Old-God who liked things as they were, but was misled by a scheming angel into thinking that totality would be even better—richer and more varied—with the projection of a newly created universe. Totality became richer and more varied, for sure—but it didn't become better. Old-God, so misled, then became the wrathful God of scripture: He created hell, introduced sin and death, cast the first deviants out of heaven, and tried to make sure that human-kind is so sufficiently beleaguered by change that God's ancient holy writs, unless obeyed, will no longer alleviate the pain of the coming of the end.

It took the intervention of the Holy Spirit, the once consigliere, now elevated into full God-hood, newly occupying the center of the "Tri-une God"—to effect a truce between spirit and flesh, to reconcile belief and practice between old and new—and promote peace between Father and Son.

Annunciation

The second theme is about the transition from the early, primordial, world to the later world—the one we inhabit. It concerns a particular (historical and ideological) event—the birth of a man who is also God. It gives rise to the unique empathy and beneficence that humans—if only sometimes—are capable of; it is also a cause of seemingly endless strife. In pacific times, this theme offers, through its symbols, an exhortation—a soft but insistent imperative for greater communion among peoples: "Fluff my feathers and I'll scratch your back." In times of strife, however, this theme carries a plea to re-humanize the enemy: to love those whom you hate in a way that will stop the fighting.

On its face, the Annunciation is miraculous—and so requires a certain pattern of belief in both the absence of original sin and the forgiveness of ongoing sin. Yet, the consequence of the Annunciation—the Immaculate Conception, the birth, crucifixion, and resurrection of Christ—augur the growth of Christianity as a social force by providing a bridge between an abstract yet punitive God, and a concrete God-as-man who offers forgiveness of sin during life, and salvation after death.

This flexibility loosened narrow ideologies and became the force behind Western civilization in its drive to power and (sometime) cultural supremacy. Christ offers redemption to all—thus giving a guarantee for life after death to those who adhere to that belief. Christianity also offers protocols for a just society—everyone, even the King, is subject to sin and forgiveness, thus giving wide credence for belief—and for waging war against unbelievers.

Within the historical span between the decline of Rome and our own great wars, the Annunciation seems to me the best point to demarcate the transfer of power from the ancient east to the modern west. After Christ, the Western world began: Martyrdoms, the Crusades, discord between the Inquisition and the Monastery, Luther's reformation, wars and more wars (most of them between Christian nations), the death of God, the secular state, and here we are—doing it all over again. It is hard to ascribe all this pain and sorrow to the (pre-Christian) God of Old. Perhaps, in his high demesnes, he did not realize—although he had created them—how skillful are these ratty descendents of the first sinners, how they could turn his noble but simple dictates into their pious but predatory purposes. Hopefully, his Son could change them—by being more like them—but still enough like Him.

End

Ends, however painful, are more picturesque than beginnings—think of the "last judgments" that were painted with such appetite in, say, the fifteenth century. This was not a question of possibility, as is an image of beginnings; rather, it was a matter of documentation—of showing what is susceptible to judgment: "Here you are—sayeth the Lord—at your end. I gave you being and free will in the beginning—and what did you do with it?" But judgments are also a way of getting endings off the hook: There was something (must have been) before creation—the aspect of nothingness that gave beginning its impetus to start—so there must be something beyond ending that gives the second nothingness its reward for having once been something.

In this sense, ends are conceptually less problematic than beginnings. In an expanding universe, the prospect of our own world's demise into a cold and lifeless cinder, responsive to gravitational forces but empty of living will, can be accounted for by predictions within physical theory—no metaphysics needed. But a disturbing question for such theory is the factor of life and sentience—especially when taken in the context of the Annunciation. This question may leave the framework of physical science untouched, but it returns us to the hoary, unfashionable, but—to my mind—unavoidable, question of the meaning of life: If life is an outcome of beginning, but will not be replicated after our world ends, then of what actual consequence are we? It is not a question that could be asked by the bacteria and salamanders, nor by the dinosours or early mammals (whose coming surprisingly took so much time)—but it is a good question to be asked by the last humans who sit on the crumbling edge before nothingness returns.

III

Divisions

THE THREE THEMES I propose here each divide in two: Creation passes from non-being into being; Annunciation enables an abstract God to become concrete by taking on material form; Judgement is the final consequence of creation, returning temporal existence into the (non-historical) realm of post-being. The sequence of themes so presented is as follows: The world begins, endures, and ends. Each of these components is identified by its own polarities: Creation transmutes between non-being and being; Annunciation embraces both celestial indifference and earthly love; and Judgment contrasts the span of contingent hope with eternal reality. Each diad contributes to the triadic form of the whole. So we have six parts—which would be three if each theme were undivided, immune to the impress of a historical dialectic, and tellable within a single context of explanation.

But we do not have this single context for each theme—for having such would negate the transitions between them, and so ignore the factor of change—interpreted variously as one of simple change, or causal change, or purposeful change. To go beyond the static categories, we reinterpret each theme in a secular context of temporal sequence—time as impersonal or time as interpreted, time as linear or dialectic, or even circular—these being various ways to identify time's function as an instrument of change.

The twofold division within each theme that I have proposed, marks the transition between the origins of the theme, and its consequences. Because these movements are transitional, they attempt broadly to impose this order (origin and consequence) on the overall scheme, but particularly on the changes that distinguish each theme from the other. The notions of beginning, middle, and end are, in fact, porous categories; they merge with, and flow from each other. In this sense they mirror their origins in the mystery of pre-being, and look for their identities as being, however achieved, in their contrast with post-being.

Before the Beginning

We imagine an emptiness which is not the same as inattention. We may be the only ones in creation to pay attention, however belatedly, to the tranformation from emptiness into meaning—and we have been trying, ever since, to find out what that meaning means. Emptiness as an image can be appropriate for where one is not, or it can apply to where nothing is not (yet) coerced into accepting what gives it substance. But nothing, even uncoerced, may not be all empty. It may be full of entities (perhaps only energies or potentialities) that vie (even there!) for the property of being something—of becoming actual. We surmise about those pre-actuals, that their chances for actuality, when played in an infinite and eternal competition, are (by lottery standards) not good. But, as we are here, there was at least one winner.

Plato, in the Timaeus, conceives of the world as being of two parts: the unchangeable, and the created. The first is the world of forms, of being, outside of time—yet it is regulatory for all that exists in the second part—the world of becoming. But forms are regulatory only to the degree that sensible creation can also be rational. The ideal measures of rationality: harmony, beauty, clarity, number, distinguish the temporal existences which exemplify these formal virtues. Such exemplification, however, remains a constant struggle—and its rejection reveals the distaff side of existence: the chaotic, excessive, intemperate nature of that (or those) untouched by exposure to the rational orderings of quantity, quality, and proportion.

As the formal order is eternal, it should metaphysically preceed—and command—the physical order. But Plato says only that the" irregular and disorderly visible sphere" must be rescued from its chaotic nature by the infusion of soul and understanding. For this transformation, Plato posits the intercession of a (monotheistic) God to effect the creation of a unitary world. This occurs through a willed division of what is unformed into the simple body demarcated by the elements of earth, air, fire and water, then into the progressive articulations of these primary elements into specific entities—the things and creatures that comprise our so-unified world.

The Timaeus presents a rich and complex account of creation, much studied—so here I note only two points: In Plato's view, order and chaos are both eternal; the point of creation being the imposition of the one upon the other so as develop the tension that creates the universe. This requires a rudimentary teleological framework—an agent (Demi-Urge) effecting this transition within a place (Receptacle) within which the world is generated. But all this is prior to our creation; it is outside the pervasive conjoining between the orderly and the good—as opposed to the chaotic–unruly. Plato gives us little revelation of divine purpose: He states that "the father of all this universe is past finding out." In this scenario, the Demi-Urge is not God, only the facilitator of inscrutable purpose—which may include disorder. Creation, in this interpretation, is the coming together of order and disorder to achieve what individually they lack: form and change, life and death, good and evil.

But this interplay presumes a dialectic that Plato's reliance on the unchanging Forms as the exemplars of value, would reject. Aristotle criticizes Plato for having a closed beginning for the universe even as they share a concept of open (unknowable but not arbitrary) ends: Plato's "being before beginning," has, for Aristotle, the fault of being timeless, undifferentiated, and purposeless. The Demi-Urge is a mystery—not a cause. The striving for completeness (mere change into progressively articulated form) that Aristotle requires, is an imposition onto chaos through his "prime unmoved mover"—whose own impetus (raising questions profoundly beyond this essay)—is also past finding out. But there is a sense of circularity in Aristotle's description of the human condition. He indicates that the unity demanded of form—whether conceptual or material— requires a return to the origins of its first impetus. Creation requires a schema for completeness—teleologic a-priori—that contrasts the soul with the material processes of change. We can move the argument here (as in Aquinas) from present conditions back to necessary causes. Whether this schema includes physical nature, or is specific only to the development of the soul raises the difficult issue of hierarchy vs. totality: Is the soul part of nature, or is it, like Plato's Forms, systematically "other?" Even within a framework of purposeful change, physical things contingently develops theough natural processes. But something must necessarily guide the contingent into an eventual rapprochement with the realm of soul. Is this power a god who begins the process then steps aside—or does God's plan, in order to succeed, require His ongoing intervention in the world's affairs—a covert (yet intrusive) teleology that continues to guide the process?

In the "De Caelo," Aristotle presents a theory of origins that supposes an act, by a power or deity, which changes (but does not create) the eternal substance of reality by giving it direction. The direction is governed by Aristotle's causal scheme—First, Efficient, Material, Final—hence, the end is unified through the necessary fullfilment of that causal scheme. But the (mute) origins of the act remain open—as does the purpose, in human terms, of its necessity.

After the End

Beginnings and endings have mysteries that the middle does not share. These are limits of knowability—not interpretation. I discuss them first so as to situate my conception of the middle within the problems of physical theory. These problems—of beginning and end—and their incremental resolutions, are to my mind, basic to the ways we want to view the middle—our historical lives—and to transform these ways into values.

Of the two cosmic parentheses that encapsulate our lives, how the world ends may be an easier subject than how it begins—endings have the advantage of being continuous (in space and time) with the historical and material world we live in. This, beginnings do not have. But ends are mysterious in their own way; we have no timeline (as we have fashioned for the Big Bang beginning) and thus little reason to believe that predictions will take us to any specific moment before the end. But, of course, we can also step away from our hubris, and admit that the end of our world is not the end of the universe—of which our world is but a tiny part. This should give the charismatics pause.

The beginning of the universe is more mysterious in the material sense, in that it asks for a reconciliation beteween two realms—those of nullity and existence. But endings are mysterious more in the pragmatic sense—for it asks how we cope with the consequences of disjunctions between increasing population and contracting resources in our world.

It took a long while after the beginning for us to generate such notions as "beginning" and "end." Our recent ancestors, the fishes with rudimentary legs, were content that they could both swim and breathe. We have finally gotten beyond this—well and good—but now we want to know what comes next. Despite the discomfort that membership in a perhaps adventitious universe may afford us, end-questions arise because of the fact that we have come to see ourselves as conscious and thus subject to answering our own questions. The ones we ask are, in the main, normative questions that address what we must do to justify our creation and duration—and, perhaps, to have our stay be longer and made better. Of course, the question also arises as to what such answers would be like and to whom directed. Living life is difficult enough without requiring us to take on a cosmic summary directed to no-one in particular.

But it is also difficult to accept that our strivings, our moral beliefs, our expectations, are no more or less than the hoppings of toads or the silent falling of trees in a distant forest. We then ask (surely we must) what the answers to such questions mean to our realms of spirit and purpose—through which we attempt to identify (and justify) the reasons for the consequentiality of our and our world's being. Burnt-out planets, after all, are no great shakes. But the physics of our world's end is not the central issue. Justification identifies concerns about our own mortality together with the goings-on in the universe. The larger issue is about what we can make of the advent of our world and the inevitable ending within it of ourselves.

An angry artist once said to me: "When we die, we rot—and that's all there is." He wanted it all—fame, money, love—before his death. But his anger does not sit too well with me. The past is not merely a parade of practical errors. There is a long history of testimonials about the notions of personal immortality and the prospect of continuity after death. But although this history can be regarded as particularly clouded by fictions, there are few among us—however bravely secular—who do not wonder at the brevity and finality of our existence, and the obdurate duration of all else. So it may be time to bring back the myths and see what they again can do for us.

We the living have long been adamant about learning how to divest ourselves from the discomforts of the void. We have lulled our philosophical anxieties by subscribing to up-beat accounts of the alternation between cultural values and practices in successive periods. It is, as the sages say, a continuing interplay between actuality, hope, and despair—but alternation is not, in itself, a teleological direction. Historical sequencing takes on the concern with finality when each progression begins to generate its rival. But prophets of a particular end-of-the-world have so far been wrong, for there is always something that emerges from the rubble: Neo-classicism gives way to Romanticism; Victorian strictures are challenged by public flaunts of free expression; Totalitarian collectivism results in the chaos of post-war tribalism. But, increasingly, the tempo seems to quicken—variants perform as if in a variety show when the audience starts to fidget. Perhaps this is the way historical change happens as it nears its end: Events move more quickly past their anticipated conclusions; cultures abandon their separate contexts and do battle within a single, global, forum. In all this, however, there seems no necessary connection between the abstract ideal of progress and its concrete reality—except that the end of the (our) world, however construed or anticipated, will come when . . .

Immortality—the transcendence of ends—if taken literally, is no less applicable to cultures than to individuals. Cultures wane and people die—and both ask for justification. Some say that in the "long run" things always get better. If we maintain our western optimism, we accept that people live longer with reasonable care—and cultures do better in climates of mutual respect. But we also know how many imbalances there are—how belligerence, easily started, becomes unstoppable, and how often ideals, attested to by all, can be ignored. We also know how well the technologies of peace feed the techniques of annihilation.

The ideal of progress and the realities of the present remain at odds in determining the social condition. Even Hegel could not show what the future stages of Geist's march would be like. The pragmatic solution—working things out as we go along—has seldom withstood the demands of saints or revolutionaries. So what can be said about the fact that no one was here when the universe began, and that as we and our world will both eventually die, no one will be where "here" used to be? A very rich man once cried from his deathbed: "Why must I (of all people) die?" The best of doctors could not answer. Anything can be said when there is no real question to be answered—and saying now about later has little weight until time-travel becomes available. The questions that I ask here, about before beginning and after ending, may indeed be questions about different worlds—which remain hypothetical, with no clear claim to existence. This is not to say that we, lolling on a large towel under the summer sun in Malibu, could claim, e.g., that consciousness creates the world. But retreating into the shade, we might say that coming up with the notion of a world beyond perception, makes our world a (sort of) fiction. What sort? Read Joyce or Proust, listen to Bach, and look again at Van Eyck or Breughel. The world we live in may indeed be that kind of world—a product of consciousness—further from reality than we may want to believe. The actual world, as an existent without us, is epistemically mute—no music of the void for anyone to hear. Nothingness, in this view, can better be described as (our) silence—the absence of questions about the world.

"Darüber muss Mann schweigen" said Wittgenstein. But are there experiences we have that, in principle, we cannot (should not) talk about? And if so, is Wittgenstein's early admonition only a thrust towards an ideal of semantic transparency—a yearning after the (extensional) infallibility between word and reference that would be ours after we have climbed up and discarded both the Purgatorial ladder of abstract logic and the semantic casualness of ordinary language? I see the "Tractatus" as Wittgenstein's counter to existential obscurantism. But his world, the scientific world that he relies on as a source for truth, changes faster than does language—and is increasingly irreverent in the matter of discriminating between observable and hypothetical (albeit plausible) worlds. So where is the "that of which we cannot speak?" Can it be accumulating somewhere else—until there are no native speakers left that must be silent? What then would be the reference of "schweigen"?

Michaelangelo and Blake did beginnings well, but they had the God-figure to instigate and move the mists along. "Nothing," as a subject of its own, comes much later than creation—it waits for the time when we have become human enough to think it as the alternative to our (and our world's) existence. The notion that nothing is eternally on both ends of contingent being, is thus a late conjecture. But it does not warrant the opposing belief in "accurate verisimilitude and true truth"—which calls for an incorrigible (purportedly timeless) account of reality, its origins and its consequences— (for goodness sake).

End as Judgment

My least favorite of the scenarios of endings is the Last Judgement—for it contrasts sin and virtue, guilt and innocence, pain and pleasure, as the overall meaning of time's passage. Judgment, on this reading, is the final end—the see-all and know-all—of creation. It encompasses the sharp edges of ancient rage and the tremulous lines of yesterday's pleading—and even promotes today's agnosticism: "I didn't mean it even if I did it, your honor—it just worked out that way." But no prevarication, not irony, neither blowsy nudity or legalistic duplicity, nor the rattling of old bones, subverts the Last Judgment's ancient claim to justice.

Absolute and final Justice (if we believe in the absence of politics in heaven) is the only factor that Saint Michael and his trusty sword must consider when he makes those last discriminations— between good and bad—and their issue, the deserving and the damned. The lesser modes of forgiveness, such as empathy, mercy, contrition, have all had their day before the final judgment comes. But is this not species discrimination? Why, I ask, when God created all there is, why it is that rocks and armadillos cannot also be subject to this judgment? It would be nice, they might intimate—those simple creatures—to be so noticed no matter what. Or are the non-judged—those who do not evidence souls—merely theater props for the more recent performances of the cosmic play?

But we must also understand that punishment—as regards the human species—is an aphrodesiac. The screams of sinners in their eternal torment are songs of pleasure and self-interest for we-the-living. You are not there yet, Joe and Jane—but just to be sure, follow the established guidelines for avoiding perdition. Even so, before you actually die, you could see a preview—just sneak in the back gate of Judgement Stadium and find a bleacher seat. What harm to watch the long-run theatre of crime and punishment being played out—that drama, if you were less timid—or more creative—might someday include you? Do you not wish—if only for a moment—that you were there for all to see—writhing and screaming in tandem with the major sinners of our world?

Another aspect of the last judgment is in its conjunction with eternity. A judicial sentence in our world to "life without parole" is pretty bad—you have only death to look forward to. But those damned souls in Hell have no death to contemplate, they have only eternity—which no correspondence course, or time off for good behavior—can fill. Instead (there is no sleep in Hell) they are incessantly goaded to emit (for all eternity) the yowls of pain that we the living listen for. So I take this vestige of belief as even more mysterious than creation. The Pains of Hell, despite the graphic glee of its multiple depictions (once a great boon for artists) is about nothing that happens. Its concoction is a tasty stew of the salacious, political, and demented.

We should note the geometry of the layers in these images: Heaven and its rewards are on top, and Hell and its torments on bottom, the world is in the middle—a well-striated top-to-toe society. But this is not merely a self-serving stasis—there is a direction buried in the verticality. Rationality of the head keeps ones thoughts on higher things; salaciousness in the groin pulls you down. Make sure then to exhibit your rationality—emanating from the furrowed brow—and then make sure your privates are mostly out of sight—behind the cod-piece or closeted by the chastity-belt. Obedience in such matters is not required for serfs, vassals, and other such; they have no grasp of purity or contrition. All that can be asked of them is that they show a proper abjectness toward their betters. Let them smell like animals while procreating, and they need not wash their feet before they die.

Note also that heavenly bliss is all-of-a-piece (even the dullest saints have their places in the choir, while damnation sorts its levels according to the type and severity of sin). There is no harmony in Hell—but lots of improvisation. Sin provides convenient categories, each personifiying an evil with its own tunes of hopelessness and rage—which may sound good to the living, but does not diminish the force of punishment for the damned. Yet, the aesthetics of agony must have an audience—all successful art forms do. But where are these sippers of pain? Look around! Are they the dispassionate souls in Heaven who know that they can have a look-see—any time during eternity—but only, of course, to reinforce their choice of purity? Or are they here on earth, wanting to satisfy their tamped–down longings for a better life through voyeuristic images of control and pain?

IV

Cosmologies

Before the Beginning

For my sense of non-being I use the notion of "condition" which is somewhat like the containment-in-no-place of Plato's "receptacle." However, I symbolize it here in Aristotle's way—as a conceptual sphere, a shape with no outside. As all internal points are equi-distant from its periphery, all distances between points are the same. "Periphery" here, stands less for "edge" than for "promise"—the promise that inside and outside will become actual demarcations within a novel condition of temporality. (But can such cosmic promises be made by no-one in particular—or is the Unmoved Mover showing up again?)

    As things stand before time arrives, the sphere of non-being does not include a factor of location in that its inside is undifferentiated from its outside because neither inside nor outside as yet exist. But metaphysics requires demarcation. So we can suggest that the sphere of non-being does have a factor that we know nothing about (like the Ding-an-Sich) but cannot avoid positing. It constitutes a promise which is evidenced by the universe that will be brought about. Otherwise, we would not be—but we are.

    We can say of the sphere of non-being that its "outside" is not determined by any particular location in its "inside." All particularities on the inside of non-being should be equidistant from its (infinitely malleable) perimiter—which leads to the surmise that our known universe, while expanding, follows the form of its progenitor in that it has no outside into which it expands—yet it does. This becomes a problem only when non-existence is recast as temporal existence—where inside and outside seem necessary conditions for the world to become—something.

My postulated sphere of non-being is large enough to contain all the possibilities that obtain within non-being while enjoying the plenitude that there is nothing (in non-being) for it to be in or on. So non-being is an aesthetically replete shapeless sphere—as there is nothing around it to define its sphericity. But we are still a long way from any other figure which could be used to show an edge-independent outside completeness. Nevertheless, I suggest—as above—that this pre-spherical sphere contains a promise—a propensity (and this is a hard theoretical part) for determinateness to be a pattern of differentiation before time. Achieving this, it can then nurture the promissory seeds for the formation of being—of actuality in time.

Because we are here, we can suppose that the nothings of pre-being had a yearning to become something. Assuming this, we also suppose that action (of some sort) must have occured as a causal factor within the sphere of non-being. But, of course, we do not know if causality is an eternal constant, or if it is specific to entities that are in time.

On the precarious surmise that time and causality have somehow invaded non-being, we might then separate non-being into pre-being (into that which has the capacity to become being), then into potential being (that which is in the running for the achievement of being), and finally into actual being (that which has won the race to be a universe). These stages, as I put them, are admittedly quite metaphorical. But I suggest to poets, artists and philosophers—and to physicists (I have no shame)—that this is fertile ground for more precise and full-blown imaginative conjecture.

Within this theory, it wouuld seem that non-being entails developments in pre-time (which may be no more static than pre-being is empty) and which leads to the transformation of pre-being into being—which, in turn, requires the incorporation (a gift from the eternal gods?) of time as a factor in being. These changes are instantiated (if we look at the theory without malice) by protrusions in the once-immaculate sphere which—through the urgency (pre-sexual, perhaps) that the entities of pre-being have to consummate their stirrings—leads to the big-bang and the advent of time and actuality. Such action could only be a consequence of something in the nothing of non-being that I call a "metaphysical yearning"—thus bringing the speculative and the actual into a first intimacy. This "yearning" implies a tendency in non-being to differentiate, even if through ongoing actions that in themselves are "innocent and pure" and thus not specific to any outcome. Such actions might nevertheless be susceptible, through the interplay between the ambitions of incompatibles and the dogmas of incommensurables (the bivalent issues of contraries and contradictories) to create variations between them that coalesce into what I call "arrays."

These variations—when "sufficiently arrayed"—first dream, then plot, then work (a first approximation of temporality) to deform the sphere of non-being's perfect nullity—its unblemished self-completeness—by creating protrusions as passages for the spelunkers of non-being that lead to being. Rites of transformation follow on the urgency for a first creative act, and they function to sift out of nothingness what can come together and create our world as an existence in space and time.

But we know very little else about our good fortune. There may be many such protrusions, each resulting in a world that may or may not have affinity with ours—that we might not yet, or never, know about. Granting this, we cannot take our space and time as cosmological realities, only as names of ways we perceive phenomena. Kant calls this pairing the "transcendental aesthetic"—thus identifying it as the underlying basis of phenomena—but not of noumena. He was careful in allowing room for the world outside of any perception—his "Ding-an-Sich" escapes experiential specification.

For our phenomenal world, space and time name its first occurrence, and so join the conceptual scheme of before and after with the extensional experience of physical reality. Anticipating a later question, it can be asked whether the nullity before the big bang is the same as the nullity on the other side of the expanding universe—that into which it expands. Can "nothing," e.g., take different forms depending on the theoretical task it plays, and are such forms open to empirical analysis—or are they, lacking such evidence, only imaginative descriptions of impossible realms? But the workings of imagination and the definitions of impossibility both change, together with actuality, over time.

Where was it then, before it (the universe) began? "There!" But that there is here—in the first sphere that follows the unnumbered sphere of of non-being. The first "then," is part of our present "now." Where, then, is the "here" of this now? It is (don't you know?) on the cusp—the "edge"—of space and time, where the happenstance of cosmic choice functions as a funnel that filters potential into actual being. Funnels, useful for recanting undrunk wine—can also serve as a symbol for the beginning of our universe. The successful array—composed of components of potential being—is primed to differentiate, and thus primed to tasting time's beginning. It then passes through the aperture. And, surely, it is followed—as in a cosmic fire-sale—by all the other arrays which, through this first occurrence, recognize their own chance for becoming real. The pressure and competition are fierce. But the image of conception is metaphorically useful here—only one sperm wins the day. On the occasion that is "The First Day," only one array passes through the aperture, and the consequence of that passing is our world.

But a world is not the universe, and a universe is not the cosmos—there may be many worlds. The universe, until recently, has been understood as the totality of worlds—of all that there is. We now accept that other arrays may compete for the creation of other worlds which may, or may not, overlap with ours—and so accept that we and they may not be part of the same universe. Recent studies in cosmology suggest that there may be universes other than ours that are (perhaps in principle) unknown to us. This is not a matter of extreme distance in the heavens, but of differences in dimension. These "multi-verses" may be universes which, while not "like" ours, are perhaps parallel to ours—in respects that are sufficient for us to exist as us and they as them, without interacting. Indeed, I may myself (and you too) be a co-inhabitant of different universes—although this co-existence is unknown to the alternative selves we inhabit in these other universes.

While the notion of parallel selves who can never know each other, is quixotic enough, the problem of totality faces a further disruption. The world has—often uncomfortably—been nestled within the infinite arms of the universe. But now that we suppose the existence of multi-verses, we need something yet more encompassing. The term 'cosmos' may be useful in denoting a further reach in our successive adventures beyond ourselves. We have some idea as to how the notion of plural universes can be approached—but we have no idea about the application of the defining term 'totality' to our findings. So let's call everything "the cosmos"—the term expands together with (everything within) all universes.

But there is always a hypothetical to challenge the completeness of any envelope—however far we may push it. Alternative protrusions out of nothingness—so the guru says—are commonplace in pre-being. Even when they fail, unlike unsuccessful sperm they do not revert into nothing—they have all of eternity to reformulate their strategies and try again. "Possible worlds" is not only a logical but a cosmological notion. We may well be late-comers in the neighborhood of being.

Then there is this: Non-being, the totality of what is in non-time, is challenged (in the abstract lists) by the realm of being-in-time—a proto-empirical upstart (like us) that credits its own existence to the imperfections in its once encompassing but now adversarial other. Imperfections in non-being, such as stasis and uniformity, can lead, even in cosmological politics, to disaffection, and so to change. As I propose, these consequences lead to deformations in the sphere of non-being. A fundamental battle between the realms of being and non-being then ensues. As time and space become the adversaries of non-being—albeit also its issue—there is a polemical need, on the part of non-being, to re-insert its characteristics into the realm of being, so that it may temper the ambitions of the expanding universe that it has (however inadvertantly) created. It would not do for being to eventually encompass non-being!

Admittedly, non-being has no spectators who can judge the temper of its status. Being, of course, has us as its spectators, who in being's later stages—our present world—have developed the ability to describe, and prescribe for, its development. The strength of being's spectators is that they also are participants in its journey. Their weakness is that they die—uncountable millions—all issues of the big indifferent bang, and yet all concerned with having a privileged place within its outcome. But they have very little time to install themselves—whether as simple memories or memorable highfliers—within that process. Only a life-time. Alternatively, the strength of non-being in this seemingly lop-sided conflict is in the supposition that being—never actually having been invited to the party—will ultimately leave. The inhabitants of our world, much concerned with the self-preservation that being teaches, may seek to emulate the expanding universe by sending their progeny to its further reaches. If that succeeds, expansion coupled with our appetite for power and reinforced by population growth (there is much room out there) will keep us going. But if this fails, if what comes to pass is that the last spectator is gone and our world becomes a cinder—and there is no sensate other anywhere to begin again—then non-being is the cosmic victor. If so, we must then—perhaps even now—deal with the penumbra (nature, meaning, purpose) that surrounds our happenstance of life (consciousness, and self-consciousness). The last inhabitants of being can then ask (why should they not?): What has it all come to? But when there is no one left to answer, the question itself becomes moot.

One way out of the finality of ending is the concept of post-being as judgement. This is frankly theological and posits an ending of human life transformed into a spiritual after-life whose disposition involves the judgment of its Creator: "How did we do ?" is now the question—it supplants the existential "what has it all come to?" But we seem not to have good answers for either question in our living. Those who want answers (really want them) may check the (soon-to-be infinite) internet, or contact the angels who attend to the rites of passage.

After Ending

In terms of time, we the people are a late occurrence in the process from the first bang-up to the last burn-out. So what can we do to invest our little stay on earth with meaning? Well, we can recognize meaning as a property of our sentience—the ways we recognize and explain the world. This gives us the mandate to invert the primacy of physical process into the primacy of mental process: Temporal and physical change both become aspects of mind's reflection. In this rarified sense, materiality is mental, and so the recognition we have of our minds' workings gives meaning to (our recognition of) the world's material beginning and end.

One problem, though, with these issues, is the tendency we have to approach them in the same way we approach physical conundrums—through reduction into their simpler elements or, conversely, by tracking them backwards in time through the increasing reach of our most powerful instruments. In the first case, we search for the elemental particles (the non-reduceable building blocks) of our universe by breaking down the smallest particles we now know until there are no more reductions we can make—even though we suspect we have not reached their end; when a larger accelerator is funded, we are off again: We break the particles we have found so far into other ones through which we satisfy conditions necessary for our physical selves. But the findings that justify these theories are not specific to them: There may be other realities—dimensions which point to realms of being that do not coincide with our theories about what exists—as in the possibility of multi-verses and dark matter.

In the second case, we seek the light of the furthest reaches of our macro-universe—which, when calculating the speed of light as distance in time, gives evidence through the emanations from the most remote galaxies, of the primordial energy as it was just after the beginning. But as the universe continues to expand, our search for the micro-second when it all began, turns into a contest between our optical advances and the rate of cosmic expansion.

Neither approach, be it collider or telescope, has given us the answer as to what it is (the ultimate particle) that undergirds the complexities of our physical life, or how it was at the simplest time of their origin (the moment of the Big Bang).

We might well suspect that these questions, posed this way, cannot be answered through the ways we search for essences or first things—there may be no such things; they may just be necessities for theoretical constructions—or fodder for ecstatic images. Continuums need not begin and end—nor need the smallest or the largest have a limit. And (to push this a bit) neither does space need to stop expanding—nor need time remain immune to variablity.

The Greek philosopher Lucretius, best known as the author of the epic piece: "De Rerum Natura," continued the Epicurean tradition of conceiving the universe as an infinitude of particles—irreduceably small atoms in random motion—whose swerves and sometime collisions, create and destroy what we know as substance—worlds upon worlds ad infinitum. Lucretius offers openness of origins without teleology. His atoms are, and have eternally been, in constant motion. Change is not end-directed but is a function of the swerves and collisions in the movement of particles. The result in a conseqent build-up of more complex entities—one of which is our world. Lucretius, of course, could not have imagined our sub-atomic universe, but the eternal nature and continual interaction of his atoms does suggest the creation of many worlds—a premonition of contemporary physical theories of multi-verses. Lucretius did not have what we would now call scientific evidence of swerving particles—except that they fit his theory of world-creation with less baggage and mystical conjecture than did those others of his time that needed an extra-worldly catalyist—a divinity—for beginnings and ends.

In Lucretius' thinking, there is no Aristotelian teleology, no journey in time toward the completion of final causes—neither is there the Platonic anxiety about the admixture of real and ideal in the proper conduct of both individual and state. Instead, Lucretius argues for the substitution of intimate and immediate pleasures that contrast and oppose the doctrinaire demands of teleological process. But this seemingly hedonistic philosophy does mask the conceptual loss—and the Epicurean sadness—of not having a correspondence between human rationality and cosmic purpose—even as its openness vindicates the drive for the satisfaction of human desires.

The epistemic pluralities that we try to stuff into a unified theory of everything, resist us by virtue of their programmatic incompleteness, and so continue to rear their bifurcating heads and confound our present attempts at closure for the future. Some say we're getting closer—but how close is closer, and is there an end in sight? The logical sorties we use against the eternal atoms of Lucretius fail to convince because of antinomies that we cannot reconcile without ignoring what may be crucial to the diversity in our own lives: We live and argue—all the while supping on the food and wine of, e.g., Forms and Particulars; Body and Soul; Past and Future; Noumena and Phenomena; Will and Idea; Mind and Brain. All good Epicurean fare. We can then take a post-prandial stroll in Kant's garden of reflection, and entertain the possibility that the world is actually one.

V

Passages

Non-being

There are many symbols for the bridge over the divide between non-being and being. My choices here are largely conjectural—even when dressed in the clothes of theories which attempt to explain what might not be a question but, rather, a performance of a question: What is the nature (interaction) of the parties in the transition between non-being and being—the transition we know as the "Big-Bang"—even though we were not there to see the opening?

My chosen symbol for non-being is a sphere that is large enough to contain all the possibilities in non-being while entirely independent of surroundings—there being nothing outside of non-being for it to be in or on. A sphere is a shape, true—yet both its completeness and its indeterminacy (one cannot tell by looking whether it spins or is at rest) pictures the conjecture that non-being is contained through a shape that separates it from something not (yet) possible—namely, imminent or actual being. Why a sphere? Well, as above, a sphere doesn't take sides. I want to avoid Euclidian angles, which are hierarchical—and instead allow for the continuity and infinity of surface that a sphere exemplifies.

An unblemished sphere (as above) may be spinning or at rest—one cannot tell. But if a sphere comes to be marked—as in our case by a protrusion—then this differential can be seen from multiple vantages. This augurs the advent of "time." As seeing is the most profound of our senses, it is also apt for personifying the first perception of the "place" of that mark—the first knowing. For us, this joins "what we know" and "what there is" into an account of how being began—the fact of existence.

Here a parodox emerges: Although non-being is both infinite and non-existent, it must have an aspect of finitude that has a "metaphysical yearning" to become something. Of course, this yearning should prefigure, yet anticipate (at least in our lexicon of the way things are) space and time. Because reality shows itself through the fact that we are here, its actions must, in some respect, be transitive between non-being and being. Taken timelessly, however, nothing can be a transitive particular within the limitless sphere that exemplifies the realm of non-being. Difficult, this. I conjecture that any particularity within that sphere is always at a center—which keeps it hidden from the guardians of non-temporality, the protectors of the constant and intransitive—which, laxly, are at other centers. This particular is the hidden worm of non-being which has the cosmic creativity to burrow through the (non-existant) walls of the (infinite) sphere.

I assume that the yearning for existence does not start in non-being and end in post-being. It is an acquired condition of both. So I speculate further that another factor in non-being is necessary for it to become an active, and transformative state. I take this to be the factor (skipping some steps) which separates non-being into pre-being and potential being. Accepting this requires yet another factor: pre-temporal temporality. This is not determinative, but it is a necessary factor for the posit of transitivity as a coherent part of my theory. Transitivity, here, is the precursor of temporality. With this pairing, I can argue that non-being is indeed divisible, and that some parts "have" the transformative (even developmental) urge that leads to the protrusions in the immaculate sphere that contains them.

It may be that the sphere of non-being was never immaculate. Is non-being ever free of its urgency to become being? Was there (once) a golden non-time for non-being? Probably so—at some part of pre-time—that part where equilibrium, and not urgency, define its steady state. But in other parts—the ghettos of non-being (here is where my notion of "metaphysical yearning" applies) the protrusions out of the steady state are instigated by an incipient and progressive disequilibrium which, given its mandate (by the Lord or chance) provides the impetus to leave the center and begin an infatuation with perimeters—which become more definite through the increasingly insistent temptations (the primordial apple) of pre-time. This, as my story goes, leads to the Big Bang—and the beginnings of our universe.

So we are here—a consequence of a something in the nothing in non-being that entails the yearning for a something else. This implies action in non-being—perhaps the action specific to many potential outcomes—that, through the attraction between similars and the repulsion between dissimilars—creates variations that coalesce into arrays capable of becoming our reality. It also suggests an anthropomorphism—for what else beside a living creature has yearnings? Well, there are computer programs and their worms; the interchange between light and dark matter; a universe with accelerating expansion.

The variations within non-being, when progressively arrayed, attain what I call a "transitive disequilibrium" which works to deform non-being's perfect nullity, its unblemished self-completeness, by creating protrusions— passages for the spelunkers of non-being to climb the ridges (and complete the rites) that point to being. Such rites are prior to holiness, yet propaedeutic to the Big Bang— and our world is its consequence— an actual existence in (what we call) space and time. But there may be many such protrusions, each resulting in possible worlds that may or may not have affinity with ours. So we cannot take "space and time" as cosmological realities, but only as names of perceived phenomena—regularities which so far suit our protocols.

Being

For our world, space and time are its first occurrence—that which joins the necessities of before and after with the traverse across expanding physical reality. Parenthetically, I must ask here whether "nothing" and "something" take different forms depending on the theoretical role they play. Quantum theory, according to, e.g., Hawkins, offers many alternatives to classical theory: Nothing and something are there considered interchangeable, each can become the other, and then reverse its status. Dark matter and dark energy comprise most of the universe, but are only ascertained as posits necessary to repair theoretical inadequacy—they have no perceptual reality. The (scriptural) interplay between dark and light may elucidate the interplay between nothing and something. Dark energy is theoretically linked in recent findings to to an ever faster expansion of the universe. This acceleration puts the furthest reaches of the universe—from which we look for data about the universe's earliest moments—beyond the penetrating power of our most sophisticated instruments.

Then too, neutrinos in recent experiments have (tentatively) been shown to travel faster than the speed of light, thereby removing the designation" C" (Plank's constant) from its central role in cosmological theory. Not surprising: The notion of a constant is always an irritant to irreverent theorists—just as the notion of God is to the foot-loose philosopher. Yet, the speed of light is what it is measured to be—even though it may have lost a race. And God may yet be recast as an integrative if not a causal phenomenon—as the player who brings harmony to, say, the multiple dimensions in string theory.

The protrusion in the sphere of non-being (the one that will be our world) appears on the cusp of the origins of space and time. This protrusion is theoretically necessary for origins in that it functions as a funnel which filters potential being into actual being. "Funnel," here, serves as an image for the energizing (efficient) cause of our world. Its issue consists of a differentiable component of potential being—which moves through the aperture, and conceivably is followed by other components that so evidence their own interests in becoming something.

But we cannot know whether this passing through necessarily results in our world. Other arrays may find other apertures and so compete for the creation of other worlds. But the profusion (and identity) of worlds is not limited to their creation. Even our own world, after its first success at becoming simple being, could have developed in other ways—resulting in different worlds. The point here is that our present world is not a necessary outcome of the conformation of its origins. Its passage through possible, potential—even actual—stages could have resulted in other worlds. But it did result in ours—so we (finally here and now) are able to question how it did so. "Possible worlds" is not only a logical but a cosmological notion. We may well be late-comers in the play of being.

But non-being need not be empty; it may have its own spectators (non-entities) who can contribute to the requirements of its status. Such spectators, although part of the play, can also be proponents of equilibrium and stasis as well as of instability and change. Being, of course, has us as its spectators, and we—as is consistent with (perhaps necessary for) being—have the capacity to understand, react, and criticize—and so further our world's development.

The strength of being's spectators (we humans) is that they also are, via consciousness, participants in its journey. But their weakness is that they all die; yet they are all issues of the Bang, and so are each concerned with attaining a privileged place within its outcome. But because humans die so readily, they have little time, very little time indeed, to install themselves—whether as simple memories or memorable high-fliers—within that process.

Alternatively, the strength of non-being in this seemingly lop-sided conflict is in the supposition that being must end. The inhabitants of our world, much concerned with the self-preservation that being teaches, may seek to emulate the expanding universe by sending their progeny to the further corners of the universe. If that suceeds, our penchant for expansion, coupled with our predeliction for population growth, will have won the day.

But if it fails, if we cannot get out of town before it comes to pass that the last of us is gone and our world becomes a cinder, then non-being is the cosmic victor—there being no-one left to attest to being.

Ending

In most discussions about states of the world, we avoid the conundrums of before beginning and after ending, and we limit ourselves to what actually is—namely, our past present and future, as they are remembered lived and anticipated, all within the framework of our experiences. This approach seems refreshing—it is indeed communal, but it has its own problems: The past is what we make it—but its secular study is now less thought of as essential for the understanding of our present than it was in our earlier sense of the past as origin and scold.

The future is a different concern. It supplants the wanted tidiness, industry, and circumspection of historical narrative with the fear that underlies unwarranted prediction. But all predictions are unwarranted—especially if we believe that our seers—from Sibyl to Futurist—can match what is embedded in our expectations. The future is not a concern in the animal mind, for "what will happen then" can be a hindrance to "what is happening now"—which can be fatal in the wild. Our own future, however, is given us through our capacity for hopes that try to balance our fears—which can be an evolutionary advantage: We calculate so as to fix the threats of future outcomes through remedial action in our present—as this is informed by our past. But as we can only imagine such threats, our present actions take on their own, often reckless, rationales—generated by skipping back and forth between our preferred anecdotes of the past, and our best projections for what is to come.

The Present

We come back to our time. The present is a most unruly state, for it vanishes the moment you acknowledge it: The present is faster at becoming past and transforming into future than our consciousness can discern stability in the state of "now." When is now? Snap your finger and you hear the echo of the passing of the present into the past. Yet, although it does not stay put, the present seems most real to us—we edit the past, and invent the future, but we live in the present. Our invented futures (there are always more than one) seek their corroborations in our past (and we also have many pasts—all of which having to do with what we were and how this joins with what we will or want to be). The present, then, is merely the facilitator—not the center—of our lives. It is where we glimpse the passing content of what, in both memory and anticipation, we call our lives. It is also, in the naive sense, where we live.

Such a present, if we slow time down a bit, can be schematically understood in its contribution to our category of "middle"—the category in which our beginnings and ends are kept as far from each other as possible. The quest for longevity is a reformulation of the middle: ("No, I'm nowhere near half way," I say to my friend who is younger by half than I). But boundaries do occur—as when beginnings falter, and we misread the road-sign of optimism and take the path that leads to the shorter end. There is no good way in the present, despite what pundits say, to look around and see where in fact we are, whether less or more than half-way through. We can believe—and most of us do—that we are less than halfway through.

VI

Birds

To FIND OTHER IMAGES for the three themes that are my subject I enlist the help of birds. You have been around long enough, my friend, to understand that themes are like birds. Hold them in your hand: The Creation, Annunciation, and Last-Judgment are alive—nesting in the communal wallow you make when you cup the left and right hands together, the birds trusting all the while (while they are still young) that you will not overly squeeze them.

The Stormy-Petrel—the Bird of Creation

This bird grows faster than the others that come after—for it has a longer way to go, and it carries the soot of pre-being upon its feathers. It is the bird of transition. In story-books, its appearance above the wind-swept sea is a sign to sailors that land is not too far away. The petrel ranges farther than do other birds, but it too needs a sometime place to sit and raise a family. Once having shown unbelievers the way to shore– it faces severe inquiry from conservatives as to why it is here at all, and how it came to know where there is something in the nothing.

Late at night, after a meal of seed and fish-heads, my special petrel answers in its birdish way: It tells me of the protrusions (it calls them wormholes) in the wall between being and non-being. These, of course, are the hypothetical holes and fictional worms which are always caught by the non-existent birds in pre-being before they get all the way through the wall-without-an outside. But one worm (of which my petrel was particularly fond) avoided the ordinary feeding ground of pre-being, and through its singular efforts completed the hole that leads to actuality. For these efforts, that special worm was, upon emergence, summarily eaten by the petrel of Creation. Our special bird, thus nourished for the first time by solid food, took wing enough to instigate the first occurrence of our world—the big-bang—and continues to this day to fly and test the limits of our universe. While the worm did not survive the bang—it was the first one of us to die—its death gave rise to actuality through its being eaten—a first predatory moment which, after some millenia, came to be called purpose.

This (as a continuation of my parable) bears on the expulsion of Adam and Eve from Paradise—which is a later, but more limited way of relinquishing the eternal for the temporal. That other worm (the apple-vendor in the tree) in this account, might be the Devil himself, or just a stand-in. (Satan's fall surely occurred before the expulsion from the garden—no?). But anyway, one result was a first defecation—the primordial shit—by the first couple on the dusty road that leads from the Garden of Eden to our world.

The Dove—the Bird of Annunciation

The dove flies mid-way in the journey between beginning and end. For its first solo appearance—the Virgin Birth of Christ—it already shows the sweetness of optimism and grace. As with doves, it does not grow very large, although its feathers are ample enough to form a communal shelter, for those who want it, from the storms it knows will come. But the Virgin's offer of shelter is centered in her plea on our behalf, the most powerful she can make—instigated, perhaps, by the pain of her own history—or assigned her, as was the agony of her only child, by the wishes of the Almighty. Yet, the Virgin's plea on our behalf is without effective power; it is limited to petition and faces the impenetrability of response. Whether her father-in-law—the Ancient of Days—pays attention to her plea is a matter of faith, not evidence. Yet, we must admit that achieving the pinnacle of intercession with an absolute God is itself a powerful thing (remember where she came from)—so the Virgin elevated becomes a consolation and refuge—a dove-song—for those who cannot any longer wrestle with the unwanted consequences of their lives.

The Raptor—the Bird of Judgment

The raptor grows slowly, but upon reaching puberty it tires of a protective hand and flies away to become what it was meant to be—a soldier in the service of divine law. The duty of this bird is to fly in ever widening circles and swoop down, indifferently, to snatch the sinners who inhabit the ornate mausoleums, the neighborhood cemeteries, the potter's grave, the unnamed ditch, and then deliver them all to the consequences that befit their sins or virtues.

Dante tells it well: The least of these malefactors—the unrepentant lustful ones—are released into the air, to be blown about like feathers by the endless winds which invite looking but never, ever, touching; onanism through eternity is their punishment. Those with advanced guilt—consonant more with the mind's more serious perversions: swindlers, crooked lawyers, and corrupt politicians—are taken up and then dropped onto fields of ice and snow, there to shiver endlessly in the cold light of a timeless day. But there are still worse sinners—perversity is without end. In response, select raptors are bred to be more leather-winged than the rest, and so less bothered by heat. These birds specialize in seeking out the most malevolent of sinners—tyrants and robber-barons, child-molesters and torturers. They circle slowly—leather being less efficient for flying than feathers—but, being heavy, they can swoop down quickly and grasp these malefactors in their talons. Despite the outraged threats, shameless wrigglings, and adament denials that their victims offer, these special raptors fly across the steaming landscape and then, without pity, drop these miscreants into the deep pits of fire and brimstone—which have recently been relocated to the right side of Hell.

The raptor of righteousness well fits the punitive needs of the Last Judgment—he is the plucker of sinful souls from their uneasy graves. Graves, according to this bird, are not a final resting place; they are the uneasy artifacts of our denial that we, the living and the dead, must all face Judgment Day. The experience of facing one's grave, be it in a pine box or granite mausolum, may be the last moments of freedom—the last chance for souls to look askance at eternity before it completes its meaning through the final assessment of reward or punishment. After judgment, there is no after—nor is there recapitulation. Memory is too private, time-bound, and too escapist for the likes of Hell. Also, even given sincere confession, memory is too irrelevant—too dangerous—to be included in the joys of Heaven. Judgment is the final form—one that recapitulates the original stasis—the one before the world began.

The raptor is the strongest yet the most simple of birds—trained to rend upon demand, but without a reason to do so of its own accord. I once had nurtured such a fledgeling in my hands, only to have it fly away without so much as a "thank-you." I am now suspicious of the raptor's psyche—as well as of its mission. This bird is much like the snarling cur next door who does not make friends even when it sees you carrying out the garbage in the same way every day. Raptors are indifferent to the cruel contrast between heavenly bliss and hellish torment; they are also impervious to the wails and supplications of the unfortunate souls they have the obligation to snatch up. Nor does this foul bird notice the difference between a plea for mercy and the gnashing of teeth—between the wails of losers and the snarls of unrepentent sinners.

The raptor is actually a dull bird: honest, without ambition for higher office, just one who takes his repetitive task, his assignments, as an official alternative to thinking or feeling. This bird, however, does have one worry: The world will someday end; the blessed and the damned will all have been deposited within their final niches—and then what will this poor bird do? We could just let him die as with other birds. But he has been a faithful part of the " big story." Let us then retire him to the prison-chaser's old-age home at the mouth of the river Styx. Those of us who are on the save-list will wish him well.

The Nightingale—an Alternative End-song to the Raptor's Croak

This is a seductive bird, enticing the individual soul through its melody to enter the communal altogether in which the seams marking individual existence dissolve within the warm need for reciprocity and communion. There is no individual sin here—it is all blamed on (and expiated by) the collective. The nightingale's song is best heard during its namesake time: night—when the residual loneliness of both flesh and spirit is most acute, and the remembered debilities of failure and guilt are most painful. This song is both distant and present; it does not include the discords of vice and virtue; and it avoids the sounds of punishment or bliss. This is because the inclusiveness of the nightingale's terrain precludes judgment of the entities it attracts; it is heard by all who look at death in their own way. These are not the squirming souls who become the subjects of the judgment that the raptor provides.

The Nightingale's song is circular in its cadences—which constantly change but go nowhere. Like music of Phillip Glass, they do not develop, conclude or climax. Such distinctions as theme and development are avoided. Within its reach the song is all the same; there is no beginning and no end—in its invocation, there is no difference between nothingness and being, or between becoming and ending.

I recommend that you sing along with the nightingale; slip-slide away, circle round-about, exchange partners—heel-and-toe or do-si-do—till the coming of the morning light. The evening's gone but never fear—it will return to you as you dream past the interruptions of your working day. When you wake again, in the middle of the night, you may not be the one you knew before. But circles always come around again, and if you wait awhile, you may meet a different version of yourself just around the bend.

Here's the rub (there is one in every scenario): If you insist on being your-self before you die, you will be accused of being version-ridden—i.e., too stiff in the categories. Like a beginning student of the violin, your bow-arm will be characterized as more proper for sawing wood than sounding notes, and your fingers will have no idea where the note you want (if you know it) is located on the finger-board. So loosen up, friend. Get used to swapping selves; learn to endure without a self (for awhile at least—we're not mean here—we won't take advantage). Forget the silly tale about beginnings and ends, and circle around with you and yours in this next go-around. Don't be upset if the you that you know is not your you, or if being and nothingness swap places like nameless loves—get used to it.

The Nightingale, who shares some characteristics with the dove of the Annunciation, namely, empathy and kindness, does not accept the dove's other characterstics—like obedience and submission. These last are linear conceits, patriarchal, eschatological—specific to the world that needs a raptor for its conclusion. In the after-world of the Nightingale, there is no punishment—except for the fault of a will-full insistence on taking advantage. The punishment for that is "shunning"—which does not inflict pain, but does demand an exclusion (for a time commensurate with the level of arrogance) from the communal rites: Every evening and in cadence with the nightingales' song, the neighbors celebrate, in song and dance, the absence of boundaries between now and then, self and other, beginning and ending, life and death.

The Owl—the Bird of the Eternal Present

This is not a metaphysical bird—but one which lives entirely in the present. The owl has extraordinary eyes which can see you in the smallest moment of your action. Then, if unimpressed, it acts like a sated shark or a busy ingenue and pays you no further notice. There are many kinds of owls, each having a present interest in the balance of nature—but each (owls are not fool-hardy) selects the appropriate prey for the fulfillment of its task. I can ask an owl: "What is the meaning of life?" and it will undoubtedly attack me—with its bright eyes gleaming—and, like a Zen Master, bite my nose to show me where living really lies. Owls live entirely in the present. They know the future is inchoate until it becomes the present, and they also know that the past is irrelevant—except to poets. Be cautious with the owl—for it is still and silent, but quick when it moves.

The Raven—the Bird that Ends in Nothing

This is the raven of Poe's "nevermore"—the bird of cynicism and the denial of hope—it contrasts most directly with the owl. The raven is a most unattractive bird; it is an intruder which, on its feathers, has accumulated the dust of the worst of history. It is a scavenger, feeding on carcasses leftover by predators—the bird that once fed on the remains of executions at London Tower. The raven does not sing, but intones its message of nihilism whenever it is asked about existence: "Death takes you nowhere but the cold, cold ground—or scatters your ashes in the river where children swim."

"Nevermore" is the polemical term for this bird; its croak invades the realities of the few who would live forever. But the Raven also laughs—a conspiratorial laugh to be sure—because it reveals the effort that is required for constant patching of the show—polishing the plaques of benevolence, piety, and charity while lowering the minimal wage. So glossed and gleaming, these worldly marks of success are taken as affirmative signs of post-life status—as evidence of God's favor, and so, as harbingers of salvation. The Raven knows better—the rich and powerful guard their present priority over the poor and meek. This dooms the least of us (and there are so many) to hell-on-earth, where we lack even the price of a shave or a bikini wax. Either way, listen for the Raven's laugh—that's what this bird does best.

VII

Ontological Questions

Now, AFTER ALL THIS, we can let the birds go on their ways—they have been helpful in separating hyperbole from fantasy. They in fact permit us, through their symbolic force, to consider other ways of getting at descriptions of worlds that extend beyond the ones we know. Birds fit well with my themes. They have an aversion to staying put; their appetites include both fruit and worms; and they have many ways of disappearing before they die. So, encouraged by their talents, I turn to the matter of making art. As with art, birds can be beautiful, ugly, or unnoticed. But humans, unlike birds, must await their representations—sometimes called art—when they themselves are done—to find out how they've done.

Art and Non-art

Some of the distinctions I have offered above can be applied to the categorical question of art and non-art. I do not question the coherence of the categories themselves—I am only concerned here with how their members are identified. In asking the question whether something is or is not art, we must also ask the question why is something "art." The answer is a matter of representing contraries (between art and non-art), and this in turn raises the issue of when the determination is being made: Is some entity, at some time, say T_1, either art or non-art? This asks, in effect, whether the entity in question possesses, at that time, the (then) requisite physical, historical, and social characteristics of an artwork. At a later time T_2, however, it may turn out that something not having these characteristics is accepted as art, by virtue of a new set of requisite characteristics which (at T_2 but not T_1)

affirms art-status. The two sets, then, conflict temporally in the task of designating "art." I take this conflict to be an ongoing chronicle of art-history. But while some new characteristics will always challenge the established category "art," the distinction between art or non-art, remains (shakily) in force. But there is no happy ending in which the category "art" will (finally) close: At a still later time, say T_3, the category of "art" itself becomes precarious through the inclusion of entities which lack the essential characteristics so far attributed to art-works, both new and old. These (putative) works engage criteria that include the denial that there is (except as a historical memento) any set of characteristics that can determine "art." Further, this radical denial questions the status and duration of the supervenient entities so far recognized—the physical object qua art-work—as co-existing in space or time. The denial of co-existence leads to such possibilities as the art-work occurring before its object, being where its object is not, or enduring past its object. These, and other such, replace the traditional notion of the actual (if not theoretical) identity of the two—art and object. The denial of a constant supervenience questions the very joining of "art-work" with a discretely appropriate object.

Historical and physical criteria can here be abandoned, but social ones remain. After all, non-belief in the status one enjoys is itself a criterion—but one which has to be constantly refurbished so as to maintain an audience for the theatre of self-denial. This requires shows of irony—plays of interchange between belief and non-belief as embellishment and balm to anxious social situations—where belief is progressively undermined, where entities which, by all previous criteria, would have been rejected are now accepted (as art)—even though some insist that they are not (fit to be) art, and others maintain that they do not really understand the word.

The comprehensive category "art" (which—as with all categories—continues in force until it becomes completely non-discriminating) must at some time, at say T_4, face the press of entities that insist are both art and non-art (both X and not X). The category "art" then faces its own demise. Its discriminating function has become incapable of distinguishing itself from non-art—which, after all, has been its historical mandate.

Category—Realm

Here, I look to change terms by replacing "category" with "realm": Categories are logically book-bound—they are subject to refutation, but the impact of such a fall often occurs after facts have rendered the category obsolete. Realms, in contrast, are more geographical—once the purview of kings, they now stake a more democratic claim to territory. All the same, they are subject to storms and pestilence—and they may be expanded or plundered. Inhabitants of realms are expendable depending on peace or war, while categories are retentive—because of their origins in language. Categories are like libraries where muted books stand side by side, hiding their uniqueness on rows of brownish shelves while waiting for a reader to offer them another chance. Realms are more unruly, subject to disturbation inside their walls and to invasion from outside. Walking in the country-side between realms can be a precarious go. Library books, in contraast, are safely sequestered by subject—if not by value. Their only joy is in being borrowed, when they are open to skimming or reading—the second hopefully more rewarding than the first—but each providing an attention the books lose when shelved again. Yet, books have no say in how long they are taken out while being read—unless their reader really likes them, pays the penalty for non-return, and so includes them in the household. Then, by virtue of their difference from the furniture and the pots and pans, they find they have become part of a realm. In realms (as one learns) every one is different, from the lord to the village simpleton, and everyone is fearful that the place they have might be taken by another—that they will be ejected with no reasons given. This never happens in a library—where, although it's dull, it's safer—except in cases of extreme rot or inattention.

It seems right to ask "where" and "when" of realms, but not of categories. In my realmic formulation, one is free to question the characteristics that members possess—and to ignore formal requirements that membership once entailed (when it was governed by categorical rules). But realms do require rulers—like a philosopher-king or tyrant—and rulers are always self-interested (unlike in categories, whose makers are, by definition, disinterested).

The questions asked of candidates for realm-inclusion go something like this: "Where were you when this or that happened—when, e.g., Pollock made his first drip?" An evasive answer is usually preferred: "I too, was thinking of dripping—but at the time I was into other things." Such evasions once identified doctrinally untrustworthy people—but these have come to be less feared than the dogmatist with intact categories: Indecision now means flexibility. When given a chance, opposing traits—like the Daltons and the Jukes—will overlap and create new and complex entities, often of surprising value. The same is true of realms. The wise ruler does not fear realm-change—commerce, not immigration, is now the overriding issue. But change, whether in the rules of governance, or in the hearts of its inhabitants, requires a rethinking of linguistic strategy—in how we talk about the things which vie for our attention. This amounts to the stylistic evolution of both political jargon and ordinary language—and of ways we consider art—our "art-speak."

Semantics of Art-Speak

The semantic weakness of the term "art or non-art" at T_1, is in its inability—in practice—to distinguish between the entities that can be included or excluded from the realm of art. Clearly, this may later become a central weakness of the new term "art and non-art"—which was fashioned at T_2, and which works to rescue exclusions from the realm of art which were due to the aesthetic criteria in force at T_1. There are always images—even in a newly permissive society—that remain reprehensible and so, unacceptable to the defenders of the realm. Examples now would be flattering portraits of Hitler and TV viewings of "snuff-films." But such prohibitions seldom last. The unacceptables at T_2 may indeed have been cast outside the walls, but they do not disappear; they are merely sequestered from the communal-aesthetic mind, and must await the later time, T_3, of curiosity about the disenfranchised past and its prohibitions—which signals another change in style and, hence, in the nature of prohibition.

This later change also signals the creation of another, more inclusive realm, at say, T_4 (around our time, in fact) that resolves the difficulty of separating the current candidates for "art and non-art" from entities that still do not fit. We can now value "reprehensibles" (The disgusting sublime—dead sharks and human feces exhibited in privileged baths of formaldehyde). These are accepted into the realm of art at T_4. But quarks, at T_4, remain outside art—not because they are reprehensible, but because they are not sufficiently present. However, as time and style and their defined realms continue to move on, "Art-solely-in-the-mind" may become acceptable as art—as long as there is hope, as with quarks, that it can someday be experienced in the flesh. Quarks (thanks to science) have become fact—and we are free to reach for them. And even such a venerable reprehensible as pornography (thanks to the Balkanization between social immorality—and aesthetic amorality) can be appreciated as a potent, if unsettling, amalgam of fact and fiction.

With the press for novelty moving ever faster in the future, at a still later time, say T_5, another realm expansion may be required—which surely would include quarks because of the amassing of sufficient evidence. But in T_5, although indirect concreteness is acceptable as empirical description, theory satisfaction is not a substitute for imagery. In this stage, "dark matter," because of its primeval taint of mystery and evil, replaces "reprehensibles" as potential for art-realm inclusion. Art then faces the problem of making an image that looks like something never seen—although, as with the Devil, much imagined. No-one knows if dark matter in fact exists, or if, like phlogiston, it is evoked merely to fill a gap in our understanding. We look forward to T6.

The realm of art operates much like academic tenure: Entities that attain the status "art-work," however subsequently challenged, are exceedingly hard to eject from the realm. Even in the halcyon days of religious fervor or individual expression, art-works that didn't fit the bill don't lose status—only value. Works, e.g., by Alma-Tadema or Bouguereau became bargains, even give-aways, in the early times of Modernism—but they remained art. Perhaps the "degenerate art" slated by totalitarian regimes for destruction, was an attempt at status change: "Crush the Capitalists in both body and soul." But many of the works so slated fled across the oceans, their bodies a bit worn but soul intact—while the totalitarianisms crumbled.

Art-works have many places for their times of recognition (as art). These differ with the various arts because of their differing formal identities as "object," "text," or "score"—and their respective modes of appreciation through "exhibition," "reading," or "performance." If we take "art" as a comprehensive category, the question "what is art" must also include the questions of "where and when is art." But to ask this we may be conjuring up a creature from the black lagoon—that is neither art nor non-art, but flourishes precisely in the unstable context that allows the intruder wannabe (for presumeably post-aesthetic reasons) to equivocate its where's and when's.

This comes about because advanced realms, increasingly a-historical, pay insufficient attention to the question "why." As artworks become amalgams of object, sound, and performance, varieties of where and when—in these fast times—are easily acceptable as criteria for the status of art. Vicarious appreciation, as inculcated through electronic media, has become a norm: Why spend money on a night-at-the-opera when you can see it on the screen. Where and when are still assumed to have concrete answers to the questions they pose—but it doesn't matter much—that's also available on the screen. But the question "why" is different; it smacks of old-time-religion; it calls for an accounting—a sufficient reason for daring to do or even possess. "Why" is dangerous, for its answer evokes responsibility, and so enters the framework of Judgment where, who knows, both you and your art acquisitions may be found wanting.

Candidates for art-status will protest that they need not comply with historically sanctioned lists of requirements in order to be art: they need not, e.g., be discrete, complete, subject to particular occasions and special contemplation; they need not be intuitive or expressive or otherwise neurotic; nor need they be socially-aware, or oblivious to, the sufferings of mankind. But a bag of negatives does not make a positive—and the question "why" asks for some.

In a sense, "why" is actually the most interesting of the four requirements, in that its satisfaction gives reasons for particular cultures to value some artworks and not others, and it reveals the criteria that these works exemplify. But "why" can also be a negative demand—as in: "Why are those ugly things being exhibited?" This direction of negative probing is generated by dissatisfaction with presently asked versions of "why," and it can lead to a historical revisionism in art-criteria. It can also lead to works that have been ignored in their time and later discovered—and to once heralded works that have recently been forgotten. More dramatically, a studied answer to "why" reveals the version of "the end of art" that comes about through a satiated indifference to the interplay between art and non-art: "They tell me the world is coming to an end—so why bother me, I won't be here so I don't care—it'll all be art."

Art as Attention

I use the term "attention" for the method of attributing status. Everything that we know as existing we know through an act of attention. What we do not attend to does not exist, until we do recognize it and include it within our world—and then it takes on the status of an actual thing (whether or not it does in fact exist).

Everything we know (as art) is known through some form of attention that generates, or is generated by, our (ever-changing) notions of art. We suppose that cave-painters had practical and magical reasons for their paintings, and that the attention paid to the hunted animals slowly gave way to the attention paid to their representations. This transfer involved a change in focus, and perhaps an incipient version of what I call aesthetic attention. In the high periods of culturally determined art, attention was codified into academic criteria that were used to identify what was worthy of the status "good art-work." The simple status "art-work" was not then in question. These criteria worked very well for long periods of time, and underlay much great art. But the criteria and the realms they served did not survive the first world war. Attention transmuted into a more personal search by artists for new reasons to create a different art that had (as yet) no clear mandate for being art.

This change marks the advent of Modernism—that compendium of styles which occurred in a time that valued attention at the expense of status. The individual artist became the arbitrator of this shift, and the emphasis on expression, dream states, non-objectivity, chance—directed attention to the areas unremarked by the older academic order where the concept of "masterpiece"—both a restrictive and enabling judgment—held sway. The attention that is required to make sense of recent art history, derives from the accelerating pace of late-modernist movements. We have not so much inherited the transformations in style as we have accepted the increasing velocity of change. Because of this accleration, attention directs itself to the social rationales of artworks—how they are noticed and talked about—rather than to any historical assessment of their qualities. In fact, many novel art-works address the irrelevance of seeking a historically grounded form of attention, by stretching the grounds of comparison. Julian Bell uses a criterion of "specificity" to present this scenario: "Courbet's "Origins of the World" invokes Van Eyck's "Marriage of Arnolfini" through the way-station of Ford Madox Brown, which then leads us to Damien Hirst. Thus multi-channelled, the new criteria can lead us to surprising corners for appreciation—and to doors that lead outside.

The collision of the content and virtuosity of beautiful art (beaux-arts) with the anti-aesthetic qualities of modern art (abstraction and action) has long been supported by the denial that form and content need be equally considered in a given work. How else but through form can one now appreciate crucifictions, flagellations and the like; how else but through content do the nudes, land-scapes, and social-protest pieces make their point. "Form is eternal, content is transient" some critics say. "There is no form without content" others say. This issue has become tired, but is always available for renewal when the work eludes location, and new distinctions as yet have no language.

New language is needed nowadays to direct attention to the emergent entity that, e.g., presents itself as both art and non-art. The specific attention required is not clear (although any attention will do for a start). It is to be looked for in the question: "Why, before this (rascally) entity is tenured (as art) cannot we perform an ontological rejection—refuse its candidacy—despite the attention paid it?" The answer is "no, we cannot"—for doing so would require social and legal changes that to art-attenders would seem both dictatorial and prescriptive—a new censorship. We cannot, in fact, deny art-status to anything that has attracted the requisite attention. How it looks or sounds or reads does not matter—neither, at this historical time, does the why. Only that the attention is paid—matters. Art-as-attention comes into focus in response, at T_5, to the claim that there is some entity that is both "art and non-art", but is "appreciable as art" (let us, until we find better ones, use the old-fashioned designations). This latitude is worth noting for future reference—say, at T6.

There are precedents: Descartes is often blamed for having constructed God through the "cogito." This seems plausible in that he argued: First, I think—which I cannot deny—and from this surety, all else follows—my existence, and even the existence of God: Deus follows Logos. But once having reversed the realms of thought and essence—by making God dependent on our knowing, the "death of God" (the Nietzchean revenge) can be anticipated through a new way of knowing, a new dogma—which operates by realigning the inherent partiality of our thinking into a metaphysical nihilism—the atheism of non-transcendence which, in turn, implies God's non-existence. Descartes "cogito" is a measure of thought's internal coherence—but not of its correspondence (with God or the world).

"Art and non-art" is a late variation on the notion of the "death of God". Rather than accepting Descartes' proof of the existence of God through the comparison of our partial powers with (our conception of) His divine perfection, Nietzsche asks that we take imperfection as a condition of thought—a characteristic of becoming—but not as a proof of an other-worldly source of perfect being. In this way we need no longer fear the consequences of God's wrath—even if we occasionally become perturbed by His absence.

Descartes died of pneumonia, brought about by the cold hospitality of Queen Christina's realm—to which he had fled because at least some of the church fathers understood the implications of his stress on mind rather than faith.

VIII

Images as Portents

DRAWING IS A GOOD way to represent the birth of the universe. Unlike its subjects, drawing is modest—it is small, often colorless, usually content with its rectangular format, and resigned to periods of exile in the archives. But because it avoids theatrics, drawing can excel in the needs of representation. Indeed , the more outlandish the subject (as creation surely is) the better does drawing capture the bones of its fantasy, and provide the form for later elaboration. Sometimes, though, the form itself is quite enough.

I am making a series of drawings that are on the theme—beginning, middle and end—that this writing is about. I chose drawing because my skills are modest and my own character is non-theatrical, and because drawings take less time than do altarpieces. No one has offered me an altar, and I avoid the thought of building one of my own somewhere in the wilderness, and then of being faced, after all that honest labor, with the fact that I do not have a God to praise or desecrate.

The Beginning

In my case, ideas precede images. I have described the originary state as a sphere with no outside, and nothing going on inside. So I draw a circle on a sheet of paper. That's to begin with. You may think that I have strayed back into the camp of my minimalist enemies—but not so. My attention is elsewhere: Neither paper nor circle are perfect, and here my task begins. Consider the paper—it has small but telling imperfections. The circle is drawn by hand and shows the minute wobbles of aging focus. But all this is to the good: A fly speck, a drop of sweat, a tremor of the hand—these will do for recasting the circle's category from nude to naked—its being to becoming.

So, by enlisting my own hand, I have found a flaw in the perfection of non-being, and I have represented it through my inability to make metaphysical perfection into a perfect representation. I take this fault as a symbol (yes, of my ineptitude) but also of the imperfections in non-being that lead to the Big-Bang. My drawings, like the world, proceed from them. It is now, for my drawings, a matter of pacing the future. How many additional, increasingly unruly, marks do I make in the circle of the second drawing? There is a long way between irritation and revolt—between the earliest stirrings and the expulsion into reality. I have to break the boundries at some point—but I cannot let it (the new-born world) just putter-around in random strokes; there is a direction to all this—whether the changes are teleogical or post-random—as befits my own transitions between styles in art. So I focus the developing matter through landscape marks—intimations of recession and horizon. Then, I insert whispers (in the middle drawings) of human nudes; and, in a development (yes) from these into variously disporting nakeds. The transition between landscape and figure is helped by walking through the forest and witnessing the sexual play—the reciprocal gestures—between trees and bushes, Heidi and Peter. The last drawings, after the cavortings of nature's fulness are satisfied, show a slowing of the battle between infinite possibility and the primordial circle. As in the sphere of non-being, the limit of representation has power over its inhabitants. My series ends as it began—with a circle—now made with a compass.

The Middle

This theme presents a different set of problems: What happened in Bethlehem and on Calvary has been described more often and more fully than has any other historical circumstance. We may dig as we will for the "historical Jesus"—and hope that these findings are, at least compatible with the storied Jesus. But this remains an academic hope—not particularly useful for the mythmakers, or for unbelievers.

The holy stories of the anguish about Old-God's absence in lived life, and the mythic stories of the Annunciation are written for those who seek resolution in the actions of a new and fecund God. In these stories, God the father, following an ancient tradition, comes to earth through a compliant angel, and impregnates a chosen human, thus welding the absolute and the contingent around a physical manifestation of both: the Living God. Shiva danced with Parvati; Zeus had a yen for the the lissome fruits of both heaven and earth; Yahweh, in the desert, was truculent, distant, and abstemious. But the Triune God's interest lay in resolving both schism and irrelevancy—through a redirection that would elevate the new tribe of Christ as God-the-Son into a catholicism of diverse peoples with sacramental assurances that the miracle of joining God and Human would be continuously renewed. Christ had to die to fulfill the miracle—to satisfy His Father. But His mother Mary's role—so antecedently miraculous—was limited to being a conduit for the "fruit of her womb" so as to effect the joining of two worlds—the old one of anxiety about a punitive emptiness, the other of a rejoicing within a new and personal certainty.

It is mainly as an image of belief that the Annunciation is important. Think first of the absurdity of the myth: Mary, a simple peasant girl, dreams of an angel sent as emmissary from God, to impregnate her—in the spiritual way—so that she might be the receptacle for the birth of the Son of God. Her conception is thus immaculate. But her pregnancy, soon noticed in the earthly way, requires that she be married to the old carpenter Joseph, patently beyond the task of impregnation but a good soul, glad to have a family—however. The marriage ceremony is marred by the intrusion of a young villager who believes that this child, this sacred concrescence of old and new, is actually his child, begotten in a farrow between the date-palms on a Sunday afternoon. He brings a stick to the ceremony, hoping to drive old Joseph and his brethren out of the temple and so reclaim the child-to-be as his own. But he is beaten back; the ceremony is completed; and the story of divine impregation faces a new enemy, Herod, who for perfectly practical reasons, wants all the new-borns killed. So the birth takes place in a manger, far from home, celebrated by the animals of the farm, and by the three kings who, needing clarity, had gotten wind (even in those old slow days) that the Messiah is being born.

Jesus was nourished on Mary's ample bosom, learned some carpentry, but was soon contesting with the rabbis at the local temple—a vocation we would attribute to high I.Q. and good genes, but one which got him into trouble—if, that is, we could call the Crucifixion a trouble. Then the western world began: Martyrdoms, the Crusades, competitions between the inquisition and the monastery, the Reformation, wars and more wars, the death of God, the secular state, and here we are.

It is hard to ascribe this to the God of old. Perhaps, in his high demesnes, he did not realize—although he had created them—how skillful are these ratty descendents of the first sinners, how they could turn his noble but simple dictates into their pious but predatory purposes. But the old God did give his minions some difficult tasks. The placement of the annunciation between beginnings and ends is one of them. In translating from the simple fable to the world-crisis it engendered—resulting in profound changes of belief and governance—one might conjecture about inscribing a different historical outcome. We could ask: What would the world be like if Jesus remained an eccentric preacher, troublesome enough (like so many others) to be hung on a cross, and then summarily forgotten? But that one time of crucifixion had a historically unique coming together of old-testament orthodoxy, Greco-Roman authority, skeptical agnosticism, and the changing demands of the Jewish Diaspora. The concept of a living god—tied "by blood" to the eternal god—one who suffers and dies because of the implications of caring for the creatures of his own creation—was powerful enough to change the world.

This is the weight born by the image of the Annunciation—not the simplicity of its myth, but rather the consequences of that myth's necessity for the changing world. There is a host of attesters to the reality of the myth, blossoming into the Mariology that had notables—kings and popes and wealthy merchants—vying to be included in the painting so that they can be seen as being in the good graces of the Virgin. These fictional images and overweening ambitions can be folded, as fragments, into a historical account—but the real theme is elsewhere: It is in the tension between the old god of punitive austerity and the new one offering the immediacy of divine love. This image is not between something and nothing; it is rather a transition—perhaps the most important in our history—between ways to understand the why of the beginning and the consequences of the end.

Take one historical account of the Virgin Mary: Much venerated but barely lettered; abrogating sexual pleasure but embracing maternal love for the one (she must have known) who would be grievously stretched on the arms of the cross. She thinks, perhaps, that this is not worth all the coming salvations. Yet, she remains passive—a crumbled heap in black beneath the cross—while those hairy fishermen: Peter, John, Mark, and Matthew—and the epileptic Paul—explain the ways of God (her son!) to man. Another strangeness of the Virgin's devotion is in her contribution to the religious rule of the celibate male. Joseph was old and passive. Whether Jesus himself was celibate, or whether he took the Magdalen as his wife, matters little. It is in the myth of the conception as immaculate, translated into the consequences of a chaste priesthood, that reveals the influence and the limits of the annunciation in subsequent history. Luther was incensed by the power and duplicity of Rome, and he also came to see that there is no need for joining theological faith with celibacy. Henry VIII wanted to become unmarried so that he could eventually have an heir—a son. He got the church of England, and Queen Elizabeth.

The End

The third theme concerns the end of the world and of our mortality, and what we can make of its coming and going. This is, in part, a theme of judgement—of reward and punishment—based on the religious narratives of why we came to be. It is also a theme of change—the scientific narrative of evolutionary development—of how we came to be. Ending is a somewhat less volatile notion than the other themes in that it can refer—even when it has other, more political, agendas—to projections which are based on the astrophysics we have so far developed—whose findings, at this point, do not deny but cannot confirm the possibility that there will be intelligent life in the universe to mark (and perhaps alleviate?) the ending of our world.

The Archangel Michael looks upon the final opening of graves, and apportions its naked scrawny and now-defenseless issue to their eternal place between the upscale glories of celestial comforts—the welcome of eternal hospitality provided by God and His minions to the faithful—or to the realm of ice and fire—the "down You Go"—to that lair of prickly demons, directed by the unceasing truculence of Satan, and the torments that can be devised and perfected through eternity.

But judgment of this sort comes late in finite being: Where was Michael in the early times? We cannot include the salamander, the granite rock or the willow tree in the play of right and wrong. We might wonder then, why it took so long—eons after the big bang—for creatures to develop who could include in themselves this (new) capacity for right and wrong, good and evil. Such capacities could not have come from the stasis of non-being—from the perfect sphere before the action began—there was nothing there distinct enough to support such judgement. We could surmise then, that the notions of good and evil are a property of late being—those times when existence, far removed from its origins, had become so complex as to need regulation—before the wars that late being generates, become a condition for killing all and every.

Or is this capacity we have to destroy ourselves, despite the suffering and the body count, a cosmic reward for having pushed being to the evolutionary point where the profligacy of evil becomes an overwhelming factor in the destruction of original goodness and the exercise of free will? If this is so, the comparison between the actual evolution of events—as opposed to a beneficent and stable celestial order—will require the cleansing that is prophesied by many—the intervention by the Old-God's second-messenger to cleanse by judging (where is Mary when we need her most?) an unredeemably corrupted world.

Or, perhaps—to speculate further—this cleansing is in the arsenal of the New-God who, being well-acquainted with the selfish habits of the old Olympians, seeks a means of continuing the myth of his own family to ensure that the distaff influences will not continue. What better way to do this than to divide the God-Head into three (a Christian Polytheism), and prepare the Young-God for an immanent return in spirit after mortal death? The first return was evidence of His immortality. The second coming (they say it is yet to come) will be evidence of continuity in the Old-God's Divine Plan, and so, proof of harmony in heaven—if not (yet) on earth.

An angry artist once said to me: "When we die, we rot—and that's all there is to it." He wanted it all before he died. But this, although true, sits shakily—even with those of us who are agnostic. He, my painter friend, is surely right about the rotting—but, with all success, he won't get what he wants before he dies. Art and life are alike in their fear of judgment—but immortality, both personal and aesthetic, depends on it. How else does art stay recognized—or you, yes you, remembered?

There is, by now, a solid tradition of rejecting the notion of a personal immortality that carries with it the consequence of salvation or damnation—St. Michael and his hairy pricklers are mostly to be found in old paintings and scary bedtime stories, beside which heaven is just a rosy cloud. The judging of art, however, continues to flourish in the still-extant world. Here, the abandonment of art-works in the garbage bin—the art-world equivalent of Hell—is the punishment for being unappreciated. Fame, shows, and sales, on the other hand, are harbingers of immortality in our life-time—signs of the Art-God's favor—as close to a Calvinist heaven as art can help us get.

Immortality, however, need not be that silly. The immanence of beginnings and the transcendence of ends—if taken literally, is no more applicable to cultures than to individuals—without, that is, finding better evidence than we have for divine intervention in daily affairs. But we need not take immortality literally. Cultures wane and people die. Yet, we often say that in the "long run" things get better. It is true, if we maintain our Western practical optimism—that people do better and live longer with reasonable care and good education; and cultures wax in climates of peace, mutual respect, and fair negotiations.

But we also know how quickly all this can be shattered. To deny any notion of progress could simply be a self-protective view of a wanting history—but it could also be darker, as it has often been—a commingling of nihilism and myths of domination that invites the slaughter of the innocent. The ideal of progress and the realities of the present have forever been at odds in determining social conditions. Even Hegel could not show what the future stages of Geist's march would be like; and "working things out as we go along"—the pragmatist's solution—has seldom withstood the demands of extremists. So what is to be thought about the fact that we and our world will both end?

Judgments, especially other-worldly ones, are a way of getting endings off the hook. As there was something before the beginning—that aspect of nothingness that gave beginning its start—so there must be something beyond ending that gives the second nothingness its reward for having once been something. The most dramatic "end-of-everything" scenario is the Last Judgement—for it comingles sin and virtue, concealed guilt and unrecognized innocence, all the accumulations of the world's self-examination after the first moment. This Judgment sees-all and knows-all; it encompasses the sharp lines of ancient rage and the soft edges of last-night's pleading. Neither blowsy nudity nor the rattling of old bones subvert its clarity. Right and wrong are the factors that Saint Michael and his sword consider when he makes those final discriminations between good and bad—between the saved and the damned.

But I ask again: What about Rocks and Armadillos and the Aspen Tree? Why are they not subject to judgment? It would be nice to be noticed no matter what. Or are the non-judged merely theater props for the cosmic play of true belief?

IX

Antecedents

Historical Art

The history of art is a better history of the world than would be a history of the world without art. This is because the history of art is all here, and now—as long, that is, as our benefactors continue to provide for the "most-comprehensive" collections that grace our largest museums. Go to the Met, the Louvre, the Hermitage, and you will see the world spread out in history in the time it takes you to walk around the galleries, tasting as you go the interests and conceits of past civilizations. Museums do not give you blame and shame; they offer the past as elevated eye-candy—lollipops for the imagination, tableaus of what you really want from life—with benches to sit on when you've had your fill. True, you have to pay a bit—but there are free days—and if you're a senior citizen, it's half price everyday.

Best of all, the world of the museum presents the larger world without oppressive chronology. The galleries readily accede to your vagrant ways. Snoop around a bit in the Old Kingdoms, then move down the hall and embrace (no touching—that gets too tacky) a Classic Aphrodite in full contra-posto. After this, go straight to the modern wing and look your fill at the splashes and squiggles, squares and smudges of a truly democratic art. Past that (unless you want to retreat) there are chambers filled with famous furniture and old-time silverware; there is a room for medieval armor and another for precious violins: Heifetz donated his Stradivarius to the Met.

On a summer day, you can sit on an outside step and watch the people—your co-appreciators—as they approach the grand façade. Watch the pretties as they carefully balance their heels while ascending the stone steps. The better heeled wear the finest coverings of silk, damask and fur—and dab themselves with expensive scents—which is the weapon of choice against masterworks that do not smell, but are nevertheless important. Underneath all that, there are problems with viewing even for the most alluring—as e.g., with Titian: How can anyone so distorted (Venus, say)—subservient as she is to old artistic whims—be so beautiful. Would you, if you could, change her proportions—or work on yours? Then, moving more quickly up the steps, you find artist-pilgrims in search of the enlightenment of greatness, hiding their earnestness under dirty dungarees or mini-skirts that show a bit of thigh.

Artists carry their own smells, for these are appropriate to expression and often convincing to patrons. When unwashed, artists can be transformed—swept into the eternal and yet remain in the present reality they are sure that art will provide. I once watched a red-haired young man looking at a Van Gogh self–portrait. He leaned unobtrusively against a pillar and focused on the painting—indifferent to the passage of time or the distractions of others. I had to believe him. I had to accept that he was not for show—he was looking at the secrets of that crazy genius for the furtherance of his own art and life.

Style in Art

Styles are confrontational—much as cultures are. Art-works compete through quicksilver shifts between love and hate. They confront each other with the intensity of their achievements—how far their portrayals have reached into the meanings that include what they—given the achievements of past masters—have not (yet) found. Contemporary artists seem to have an advantage here—they can look back and choose their adversary, as if from the selection in a candy-store. But this can work against them too, as the distillations of past genius seldom carry the uncertainties in vocation of its own times. The fumblers in the present, then, succumb to the pessimisms of their present: Although they learn much about coping from the past, they suffer from a terminal ambivalence about it—about their lack of the skills the old ones had, and their lack of the good-reasons these old ones had to learn those skills.

But settle back, friend, and don't try to heal uncertainty by escaping to another time; choose your heroes carefully for the realities you now face. Comparisons across the centuries can be vitiating—they make it easier to give it all up before you get high enough to kick the ladder away. So when you choose, choose the most fearsome adversary you admire, and be content with a good fight.

In so choosing, one must avoid both the constrictions of genre, and the oppression of strict chronology. This is difficult: In our permissive times, any works that one might choose, if sufficiently stroked, could be given a place within a sequence, and aligned with any other. Multiple interpretations often generate unworkable working methods. Dissident interpretations that deny the very art they appropriate are everywhere around. Works unwilling to face the enormities of past art, may back away and call for historical inclusion by deconstructing history.

Beginnings pose different problems than do ends for my thesis—they move from nothing to something. But as making images is human stuff, there are few extant ones that take us back very far. The oldest ones, like Cave paintings or Egyptian Hieroglyphs, are about the beginnings of something already present—but not of anything before. No paintings I know of are about God's pre-universe dilemma, or even His pre-Adam and Eve anxieties. But there could be efforts—especially now, considering the reach of the electronic world—that take up the changes from before the primeval world began to its inception—the indeterminate play of particles that transforms into the specific consequences of a first material happening. The next scenario would then probe into the first and most famous couple's dilemma—the choice between apple-of-snake (life and sex), or the eternity of celibate calm. This choosing—distinctly human in character—marks the beginning of time and historical change. Given the richness of those subjects, there should indeed have been more art that focuses on the state of God before the advent of creation and the expulsion from the garden—but there is very little. It may be that the slow road from early vegetation through late dinosaurs—absent luscious maidens or exemplary beheadings—is not that interesting.

I imagine a sequence of "possible-to-actual worlds" by first bringing together the early abstractions of Philip Guston and Cy Twombly, together with pre-historic cave–paintings and the artworks of precocious six-month olds. These wisps, drips, and scribbles, although separated by millennia, all show the non-directed bubbling of incipient pre-beginning. The gap between bare beginning, and the beginning of something, can be filled with the yards of Monet's "water-lilies" and Pollock's "drip-paintings." To stretch this point, I also include a Mondrian, the "plus-and-minus" paintings that show the modernist anxiety about flatness and perspective—the distinction between a painted abstraction and the painting of a wharf. These all indicate ways of showing what would be primary in the representation of beginnings.

I now look at the 15th century transformations that I identify as the "beginning of the middle." These early Renaissance works, herald the escape from schematic formats through their replacement with the affective subjectivity of the Annunciation. To buttress this transition from beginning to middle, I enlist the "Baptism of Christ" by Piero della Francesca, which brings together a lingering medieval abstractness with a recovered classic naturalism. Birth is an apt symbol for this time—the birth and rebirth of Christ, the rediscovery of Greek philosophy, the redemption in art of the transient persons that symbolize eternal ideals, and the re-emergence of argument between the sacred and profane.

I continue on with Raphael's "School of Athens" which brings two sages—one older than the other—together in a polemical walk, through a noble arch, past lesser minds, past even a nude or two, as they argue the fine points of life's meaning. Can anyone doubt that they will find good answers? I earlier had thought that the two were Plato and Aristotle, but scholars say that Plato's adversary is really Euclid. That make some sense—given the Greek love of geometry—but it does not open to the grand contrast between St. Augustine and St. Thomas—between the intensity and fragility of belief, and the rigors of reconciling God and flesh through logic—a contrast which has bedeviled and energized the faith for a good while after. For that reason, I prefer my pairing. Raphael, himself, did not identify his depicted adversaries.

There are other exemplary works for this stage—Leonardo's Last Supper (which shows what one can know about the unspoken delicacy of that event), and Michelangelo's Sistine Ceiling in Rome (which shows everything that one could imagine about the urgency and necessity of the beginning). But a day's journey takes you from Florence to Venice; and in that short distance, much changes. Florence is the home for the religious works of Michelangelo and Raphael; Venice is where the Venuses of Titian, Tintoretto, and Veronese, were painted. These latter works do not directly contest the importance of religious concerns, but they do show, through powerful figures of a beauty found elsewhere only in Classic Greek sculpture, the advantages (and continuity) of pagan delights. This contrast, between the Christian sacred and the Pagan profane, creates a space in Christendom and divides it between the private and the public image—a space which, throughout Western history, has been used to enhance the categorical distinction between power and person. But the Venetian paintings, although artifacts for the delectation of power and wealth, do question what the powerful had earlier believed—as formed by the teachings of Romanesque and Gothic art—and show what they became willing to entertain: representations of Pagan sexuality side-by-side with Christian images and beliefs.

antecedents 55

To continue on to the center of my narrative's "middle," I select two paintings: a Rembrandt self-portrait, and Rubens' "Wedding of Earth and Sea." The undisguised glorification of sexual images in the Rubens provides a contrast with Rembrandt's probing into deep levels of subjective insight. This pairing foreshadows the conflict between objective form and introspective need that becomes a staple of the twentieth century avant-garde. Both ways are increasingly manifest as the middle ripens, and both capture the intensity of the conflicts that follow from the Annunciation: The subjective questioning of the source and possibility of faith, and the public concern with the power of belief and its public manifestations.

The work I choose as the great conciliation of the middle with its aftermath, is Las Meninas of Velasquez. Power, propriety, and skill are upheld, but pictorial logic is torn asunder—in ways more subtle conceptually than in any other Baroque work, and certainly richer in pictorial demonstration of multiple meaning than found in most modern works. The painting presents the child—the queen to be—surrounded by attendants (Las Meninas) two dwarfs and a dog. On the left, looking out (at us?) is Velasquez busy at a painting we do not see. He, in fact is not looking at us—but at the king and queen we also do not see—but who (like us?) are ensconced in the space outside the canvas, and whose portrait he presumably is painting. The tableau of the painting (the Infanta and her servants) is presented as a diversion from the tediousness (even for royalty) of posing. These Royals inhabit the space outside the painting that we thought we were in—as Velasquez pictures himself as looking out at them, and at us as well. But there is another (outside) space that the King and Queen are in. We know this because of a minute reflection of the royal couple in a small mirror at the upper left of the painting—showing the space that we, as observers, are not in. We have been supplanted—removed from participation—by the Baroque representation of Spain's royal couple in their palace. To complete the drama, there is a cavalier, pictured at the upper right of the painting—sword at side, the exemplar of absolute power, and constant guardian of the royal family. He looks dispassionately at all of us. There is no subjectivity here, as Rembrandt would offer. Instead, there is the affirmation of the future by an already vanishing time—by using the extant style to give us incipient visions of what is to come. This is a secular painting—no hint of divine presence. Seen one way, there is a perfect harmony: royalty, privilege, servants and retainers, a few grotesques for amusement—a self-sufficient and unchanging world. But the painting also offers an astounding premonition through its multiple interpretations, of the ending of that world.

For the late middle (a beginning of Modernism) I choose Courbet's "Studio" and contrast it with the Velasquez. In the "Studio" there is no longer a cocoon of monarchy to provide the wherewithal for the exercise of genius. The Courbet is powerful as a painting, but it is also poignant in non-aesthetic ways; it is about sadness, indifference, and futility. These are social ills—and the painting is a social critique. The canvas is divided into three tableaus. The left side is devoted to social misfits, to the impoverished, malformed, and otherwise unwanted. On the right side are the proponents of culture—poets and philosophers—also unwanted, but nevertheless feared for their influence on the revolutionary potential of the misfits. Both sides are painted in grey tones, and the ineffectual figures in both stand passively looking out—not at us, but at the world they have come to share. The downtrodden need a plan of opposition to their plight; the intellectual, by all rights, should give society sufficient reason for saving the downtrodden—but, as Courbet knows, this does not happen.

The center tableau is of an artist sitting at his easel, vigorously painting a landscape. He shows no interest in his visitors, and his large and very naked model stands behind him watching as he aggressively limns the leaves and sky—thus relegating the poverty and the intellectual impotence of the two side groups—even the sexual charms of the hovering model—into irrelevance. The value of an art that is about itself now becomes the primary concern. This is not Courbet's self-portrait—although he uses an aspect of himself (his back) as a model for the artist. Rather, it is a parable of an art that has lost its traditional subjects, but now has another instrument—the disengaged artist—for finding a new subject that is impervious to the difficult issues of social and political—and even personal—life. Courbet's parable is a harbinger of the Impressionist move to an innocent subject—nature—and it thereby points one way to Abstract Art.

Here, my narrative moves to the beginning of the end. This is the time just before ours—although still one that some of us were part of. As a contrast to Courbet's melancholy, I offer a version of Picasso's "Three Musicians." This richly patterned and intricately ordered work is, on all critical counts, quite abstract—no atmosphere, no perspective, no light and shade. The figures are clearly evident, but as patterns they remain part of the abstraction. There are no life issues here—no more than in the landscapes of Cezanne and Monet. But Picasso himself bites into the indifference of Cubism toward its subject. This is a city art—harsh and clever—the glee in the abstracted faces of the musicians is directed to (is a criticism of?) the consumers of Modernism. It also counters the Impressionist ideal of nature's purity and relevance.

The American (post-war) reaction to the European artistic rationalisms was one of ambivalence—a version of the "unhappy consciousness." The artist need not (cannot—for art's sake) probe into the normative implications of what is now his subject, but one will love and hate the subjects one cannot control. DeKooning's "Women Series" is a good counter to the Picassos "Three Musicians." The manic, mechanical, glee on the Picasso faces, confirms their subservience to the abstract subject. But Picasso let this interplay—between pattern and representation—remain in these works. They are historically, a confirmation of his rejection of formalist dogma—and of his unwillingness to abandon illusion. For deKooning, the formal debt of Cubism is undeniable, but the subject of his "Woman" series is of a coital rage that combines the overt geometry of abstraction—the tough drawing—with grimaces, splashes, and disfigurements worthy of Goya.

But rage becomes boring when times are good—and times did get better in post-war America. To liven things up, then, a new rapprochement was in order—between that which gave us all those many emblems of angst and radical form, and that which would erase the difference between high and low in favor of an aesthetic that feeds the enjoyment of what social affability, and titillation supported by money, can bring.

The practices of co-optation between high and popular art often provide art with a new vitality—but they are also, for me, a sign of the end of the middle and the beginning of the end. Warhol's "Marilyn Monroe" is an exemplar of one way to get past the middle and skip to the end. One aspect of Warhol's "popular" interpretations is his sapping out of all prurience, all singularity, from the person who is its image. "Marilyn" is given us in multiples—distinguished by color schemes. We can prefer a magenta, pink, or green version of the goddess—to fit our every taste. There is no impossible demand, no heroic feat left to gain her favor—no anger at the unattainable, nothing left even for voyeurism—no forbidden fruits to endanger the path, now newly paved, that could bring her here to you and me. Warhol has done his male-gaze bullies one better. After he did his graphic dance with Marilyn, there was no way for them to remember what she once had or what it was we wanted. We remember very well what deKooning's "Woman" had—but we didn't want that either.

In the history of Modernism—the later portion of my thematic scheme—there are other versions of the artistic retreat from the worlds of pride and fear. The alternative to popular art—whether Action–Painting or Pop-Art—does not always end in subjective thrashing or affective blankness; it can also be evidenced by a late-monastic introspection. This is a mark of the effort to deny the demands of the expanding middle by not feeding its appetites for entertainment—but rather to purify it by offering a new and difficult subject for contemplation. Works of this nature can be found in Malevich, Rothko, and Ad Reinhardt.

For historical reasons, the anguish of Malevich remains public. As we know, he was a victim of a totalitarian ideology. The fall from young optimism towards a nadir within which historical forms no longer could be built, was for him (and still is for us) historically painful. The once-cherished verities—the act of drawing into an emerging subject and the rewards in virtuosity—especially by remaking the masters into new and personal achievements—this, for Malevich, at a crucial point, lost it's possibility. The tutorial past and the novel response are now both prohibited—only totalitarian kitsch remains. To avoid the death of a complete silence, Malevich offers a white square (tilted, and so, programmatically insecure) on a white ground. The stance here is not of disinterested contemplation (as Ad Reinhardt would later want for his own end-game) but an exposure of emptiness—and the sadness that futility brings.

Reinhardt, despite the persona of an outsider scold that he cultivated, is in comparison, a child of the Enlightenment. His end-image is geometric, symmetrical, and black on black; his view of ending is dispassionate and fatalistic, sometimes admonitory—but always serene. His antecedents are in his admiration for the timeless, repetitive, impersonal art of ancient Buddhism—and its transformation, by such an obdurate soul as he, into the affirmation that continuity in art is (must be) separate from the call for novelty in art—although pervious to the changing forms of life.

Rothko also forsook the game of artistic alienation, but he dove as deeply into the present as his sensibility and anguish would allow—hoping to find there a consonance with a religious concern that transcends, but does not exclude, the personal. Through the oppressive yet beguiling play of shifting color bands, Rothko climbed the ladder of his otherness into the late black paintings—still suffused with the incomplete covering of their underlying colors. The reactive quiver of the flesh underneath the blackness cries for our attention. This is not decoration: it is passion. It is also Rothko's defense against the impersonal image of ending—the image in which passion would be supplanted by indifference as the remaining way of life.

Reinhardt's end-image—conceptually blacker than Rothko's, but physically not as black—is not at all supplicating. His images, symmetrical and barely discernable, hide nothing—they reveal nothing either—but they evoke in their calmness, another way of meeting the end—which, in fact, may have always been there, waiting for a proper recognition. Reinhardt's art is not a personal geometry; it is as close to the depiction of an ideal as can be pictorially approached—one, if believed in, would clearly show the end of art. We could then all rejoice—and return to the nothing from which we came.

With such symbolic support, the end of the world should have happened then—at the close of the first half of the 20th century—when all the antinomies were festering in the open, and signs of the coming trivialization had just become evident. But what did happen was only the end of art—not as a killed enterprise, with Warhol trying to take up Duchamp's sly dirk—but as the decline of seriousness, the relinquishing of shared responsibility between symbolic representation and the character of actual life.

But if we evade the importance of such an end-game, diminish the seriousness of aesthetic minimalism and the consequent slide of art into epistemic futility, how the end should look takes on a casual, a-historical, although ambitious, face: Anything goes because we're not going anywhere, and so we're free to do anything that we at this late date (it's not too late, is it?) can think of doing. But the conceptual proximity of ending, like the recognition that one is old, delimits the abandon—the Dada impulse—that once showed the variegated pathways to enjoying full futility in our lifetime. Now, having outgrown such heroics, we must (some of us) grind on to complete what we have started—and not think too much about another man's wife.

X
Preamble to Making Art

ONE VALUE I REQUIRE for a contemplation of making art is the freedom from inclusiveness. We have now amassed too much art history to be clear about the decisions needed to make the best and worst of art. Appreciation is no good for this—art is not a making, but a way to pain and pleasure—which is only a way to what can be done.

Titian had his Giorgione, and Picasso his Cezanne—the love-jealousy between them energized the effort, and the ideological need for a new tradition directed it. But despite today's incessant need for a novelty that is independent of tradition, there are things in the past to be looked at—if only for another way to get from there to here. Remember the times when the "here" had become empty, and yet was met with accusations of blasphemy or treason? Our newer changes, in contrast, are not something to be loved or feared (nothing in art is that disturbing any more). What sacrilege or immorality could now so exercise the polis that it would be banned? Certainly nothing that attracts profit and attention. Of course, it is a good thing, this freedom—this lifting of restraints from what can be shown or said—we could not want it otherwise.

But there still is, for me, a question that does not go away: Is there something that has now left the art that once served the monarchies of church and state—the art that rose to pinnacles of greatness while glorifying the narcissism of absolute rulers and the propaganda needs of autocratic religions? To reify the gained historical freedoms, Modernism took on the Impressionist images of innocent nature and later the subjective images of Expressionism in order to lift the arts away from collusion (or confrontation) with the powers that needed entertainment. But the question that rankles is whether art, in taking these steps toward its aesthetic freedom did not also move toward irrelevancy—so that the solution was to return art to new domains of power—now exemplified by galleries, museums and the internet. What could artists show, in this all-permissive jail, but works that are edifying for the social weal without the need for personal commitment—works that show a willingness to join with (if asked) the power of the enterprise "art-as-art-is-fun."

Sometimes, as in my story, rejections require a long journey. Along my own way, I became antagonistic to myself. I had once rejected girls with hairy legs—not because of the hair—I like hair—but because of the skinny-pointy-angry polemics that anointed them. Later, I discarded my art pieties by considering them as the fat-headed beliefs that had supported my—and all those paint-smeared others'—adherence to the style that, as we were told, would last a thousand years. With the girls who didn't shave it was easy; I just walked away—they didn't miss me. With the art, it was harder. I wanted to leave, but the carousel was still moving. I had to think of myself as free even from my own ambitions before I could risk jumping off.

I do not remember exactly when my pieties began—why I first got on the carousel. But I do remember that before I jumped we were all together for a while—the rest looking just like me. We all wore the paint-spattered dungarees and the dirty blue shirts of freedom. But then there was that time—an ordinary night in winter—when I saw them circling around on 57th street—on the Tuesday night of openings—chanting what sounded like the God-anointed incantations of my childhood. They all had big heads and smelly whisky smiles. If we stay together, they said, we'll make it all together. When they weren't watching, I ran like hell.

XI

Proposals for Making Art

HOWEVER UNFATHONABLE THE REASONS, creation actually did happen—and as we are its issue, its way is a paradigm for how we began—and it is also a model for how we can make an art of how it happened. You see, pre-creation is mimicked by pro-creation.

I want you to contemplate an artwork (not yet made) that does justice to the themes I present here. This, of course, is puffery. But by my accounting, I live in the latter portion of the late middle where puffery, e.g., preening, and its companion, irony, e.g., dissimulating—are the best rejoinders to an out-of-focus world. When, where, and how I live fits my given span, my present interests, and my dissatisfaction with my past efforts to do what the old ones indicate "should be done." So, when asked, I merely propose the art I have in mind. That art addresses the three themes I here identify, looks to their differing symbolic needs, and attempts to show them as being cogently together in one sequence. But as the sequence is cosmological, the themes hypothetical, and the works themselves "in-process," my attempts are not yet for all to see.

A First Proposal

Here, I propose a painterly account of the transition: We go from nothing to something (the mark as first instantiation) then to the extension of mark into stroke (the first response) then the congealing of stroke into shape (the first formalism) then the transformation of shape into image (the first representation) then the specification of image through a format (cave drawing, altarpiece, easel-painting, computor image). In our world, formats are the factotums of the process of creation. They identify and contain the images of a some-thing that comes from a no-thing. Marks, strokes, shapes, and images are properties of potentiality. Embodied in formats (styles of containment), they are the ways of getting art to symbolically cohabit with the stuff of actuality.

But while images are usually of things, their compliants (unless you specify them with, e.g., a title) indifferently occur in, on, through, from, behind, before and after, each other. Such compliants, in sequence, comprise all of history—everything we attend to. But the compliant of an image may, in fact, not be a thing—but rather an imposter, a facsimile. Its presentation may be a masquerade—only a chimera slipping through the lexicon of determinate being (think of all the wet nudes in the caves of fevered imagination). Images can also be about nothing—but, themselves, not merely nothing: to be images, they must contain some variant of mark-stroke-shape. Some such images can present themselves as being about themselves. They can be newcomers to this gambit, as occurs in abstract art, or old first-timers, as evidenced in geometric cave-painting.

One can insist, of course, that these "self-referentials"—from the first hieroglyphs to Judd's anodyzed cubes—all make reference to compliants of some sort. This is true—but not startlingly so—it merely affirms that everything we take note of (attend to) is an image of another thing; everything in the world can be both symbol and compliant. Abstraction is a particular way of referring—it is a style, not a new discovery of how the world is, but a new way of ascribing artistic value to images that heretofore were regarded as symptoms of illness, signs of spillage, or "mere things."

If I were a scold, I would insist that images of being and non-being (for those who want to play the cosmic game) should be tempered by reflections on the origins and demise (the historical perameters) of what gives rise to images. Then we could insist on certain differences between image and actuality: If we were art-political, we would valorize those images that are presented for present purposes. But such aggressive images, although intent on conquering, changing (or, at least, explaining) the world through their insights, still want to remain art because of their originality—as opposed to more prosaic ways of explanation.

I am not a scold, however, and I continue to like the interplay between the slippery inhabitants of illusion and the stolid ones of reality. The first do not stand still long enough to be catalogued—and they thrive on the changes that all the arguments about their finality provoke; the second provoke through their programmatic insistence—although they tend to be formally incomplete, and soon irrelevant.

We may try, if we are otherwise ambitious, to extend the self beyond the time-inflected world—where we could sit on a privileged cloud and observe a reality that joins the ending of things with the first signs of their origin. We could then know how something comes to be both other and the same as what it is—how it is that things that are, can also be images of things that they themselves are not. Representation has an ongoing battle with actuality (although, for us, there can be no actuality without representation). This is one reason for making history.

The art-work I contemplate (with its magisterial subject) could refer to more traditional periods when representation was less questioned—it could in format be a triptych, one that combines its divisions into one great panorama, and yet holds the smaller concerns within discrete sections. Such a triptych could be about some ideal past, a preferred present, and a wanted future—but not the ones that mark the times of my life. This, paradoxically, could give me some leeway: I wouldn't have to care about myself. I could stretch my representational skills and make three paintings, one for each theme—"beginning," "middle," "end"—but on a single huge canvas (of course, requiring scaffolding and nimble assistants) so as to insure that they (the themes) although separate, remain together. This makes it harder for critics with scissors to find the thematic boundaries and then decide whether, despite the intentional format, they look better when apart. But I most probably will not take up such an approach—mostly because I cannot paint as well as, say, Tintoretto or Rubens. Perhaps I'll reconsider if I'm offered a large public wall—and an old master's stipend.

A Second Proposal

No, I won't reconsider—too much work to go back there. But I could take up one of a number of current styles: I could, e.g., go native and strew found objects around the house and garden, hang moldy memorabilia from exposed rafters, bang on faux-skin drums, and populate my garden with coral-snake look-alikes. I could then deck the halls with TV sets playing loops of sound and sight—which were recorded in the earlier flagrant-fragrant days. The distribution of leaflets that encourage participatory enjoyment in the nightly play of themes is both fiscally and theatrically sound. (I plan to charge admission and sell drinks). But the fine print in the leaflets would warn the unwary about the un-natural demands and moral hazards of my subject. There is no necessary correlation between bad behavior and bad art.

A Third Proposal

I won't do that either—too success-prone. So after a time of unfruitful scratching and scowling, looking for other ways, I have come to this: I will take (some-of) the nothing that is outside the world of art, and make it into something that is art by re-assembling what is outside—abandoned artifacts, last-night's garbage, lost waifs and weary revelers—into clusters where I can join the things I have with those that have tumbled down my dirty staircase. Then, standing up slowly—I often tumble too—I will push, sweep, rake the manifold that I have gathered right out into the wet streets, at rush hour, to face the oncoming traffic. These once modest and separate images, the ones I had formerly brought together on a single canvas, achieve a new attention. They now include the traffic jams, wind and rain, oppressed pedestrians—together with my own willy-nillys. My latest strategy requires that this, my novel art, be substantial—not prissy canvases or precious works-on-paper, but sheet-rock remnants and bug-eaten driftwood mixed with ancient stones from an earlier obsession, topped with solemn and portentous books (I have a lot of those), empty wine bottles with labels left on, scraps from last night's dinner, shreds and tatters of underwear left over from lost loves.

Additionally, for the opening (the show is only one-day long) I have asked some special people to become a part of the art—these are people I have found who prefer being part of something not needing explanation. Sometimes they take their clothes off—sometimes not.

Out onto the streets you go my dears—all together—an admixture of what "was-art" with what "will–be-art," brought together by descendants of the worm that met a first-ever imperative by burrowing through the wall of non-being. It's coming time for the opening: The traffic is heavy, and the streets are wet. There will be accidents of course—the twisted metal we all admire is also the medium of skids, wrong turns, and deadly vulgarity. Cars, especially when driven, don't understand art—they prefer shattered glass and screaming blaspheming sprawling victims, with bits and pieces of the way-to-go, as topping. But this is all part of the art.

Disclaimer

For those who wish to stay on the pavement and argue that this mess is not art—only the consequence of post-historical ignorance (a perversion of the avant-garde) I have deep sympathy. But I have committed myself (who else can commit me now) to advance (at least in theory) a deviant approach.

But I want no cleanups, lawsuits, or old-mentor mutterings to interrupt my efforts. The pedestrians who stumble through the rubbish will, I know, object: How can someone be so destructive to life and property—to our welfare, comfort, and safety—all in the name of art? Of course, they have that right—life is more important than art, we all agree—but they are loud and some of them are ugly. So I will go back upstairs—even though there is nothing there.

XII
Perplexities in Making Art

About Images

What an image is that is also art—is a long story. Like other images, it answers to "what, when, where, and why." All images stand for what is imaged. "Stand for" is not a substitution but, rather, a reference that defines the subject in a way other than how it is (was, will-be) when it is "stood for" at another place or time: A shoe may evoke an image of comfort and ostentation—but Van Gogh's shoe is about the travails of a people. The value of images is that one never knows what, in fact, they stand for—for the reason that there is no "in-fact" when it comes to imaging. The destruction of the autonomy of their subject is what images do—they show their subject in certain ways and not others—thus depriving it of its (ideally) full reality.

Plato's "Forms" are not images—but all their particulars are. For a particular to be an image, one must pay it a certain attention—governed by place, time, and context—in order to distinguish it from other ways of its being attended to. This done, and the myth of "correctness" denied, images become anything specified in a given mode of identity the mode in which attention is paid. Some images, under a certain attention, can be true others can be art—some both. But none need be so, or if so, need not so remain. Modes of attention change together with the images they produce. There are the histories that codify these changes—until they themselves are rewritten. If you hesitate at going this far, look around, and try to find something that you attend to in a certain way that is not also an image of something else—under a different attention.

A First Concern

The last and putatively the wisest spectators who witness the coming of ending, may deem their own ends as affirmations of the universal belief that we have been the most attractive of the many resolutions (possible and actual) of the Big Bang. But this, of course, is an a-posteriori judgment—the singularity that produced us happened before good taste began—before time in fact—so there is no way to know which others (the fruits of other Bangs) would score higher on the gauge of cosmic beauty. We could ask Venus, who knows about such things—but she is old and somewhat out of it. So we are left with the uncertain residue of what we—artists and critics, philosophers and pedants, have come to think about such things.

But is there art and beauty to be found in endings? Are there also protrusions in ends that can return us to early innocence (revealing long-dead aunts with pendulous breasts and warm smiles) that will mollify the inexorability of times-ahead (the coming attractions of haughty anorexics on nine-inch heels)?

But another protrusion demurs: This ambition to translate existence into art requires more than one reality—which we do not need when we play tennis. It also requires a metaphysical transparency of categories—such as between art and non-art—which we need, but do not (yet) have. Further, it proposes an intimacy between pre-and post-history—as before the Bang and after the Burnout—that we, as time-and-space-bound existents, cannot have.

A Second Concern

Consider the program: "All-the-world-as-art." Yes, it raises many questions: What happened to non-art? Is non-art the dark matter—the stuff we do not perceive but cannot live without—in order for us to be art or to be at all? If so, if we want our life to be art, our "being" cannot be the ordinary life we just happen to be living at the moment. "Art-as-life," then, must be more than simply the continuation of the death of "art-as-art." Consider further: Does art-as-life follow upon, or is it synonymous with the death-of-art, or, like dark matter, has it always been there?

Our aesthetic ambitions seem to be thwarted by such semantic muddles: "What is art" is only surpassed by "what is life" as an ongoing irritant. As the term 'art' is a social designator which has flirted with all of truth, goodness, beauty, sex, money, and power in its checkered history—we would never get substantive agreement from the polis on its appropriateness for any coming style-of-life (such as, e.g., modeling sensibility on a scale of right proportion—or basing civic legislation upon the aesthetic value of expression). Given our history of wars and other such relics of the waning days, there are too many despoiled stretches in our world for it, on any count, to all be claimed as art—however giddy we may be at those times when the shooting stops.

A Third Concern

As I have made nothing yet, there is still time to propose a work that is smaller in scope, leaves most parts of the world untouched, yet is subtler in ambition than the ones I describe above. I call it: "Some of us can be art." Under this new proposal, our becoming art is optional—there is only the penalty of cosmic lack of interest (by those patrons who inhabit other cities or come from other Bangs). However, this proffered work (us) does have some virtues: It does not flirt with timelessness; its content changes with the speed of the information grid; and its images are fashioned to describe the inside and outside of what we at any moment (think we) are. Because of the many locations of its content, this work—unlike the inhabitants of an all-embracing, eternal art—is never at the center of the world. It is a democratic art, variable in scope, and so requires interplay—a game, if you will—whose object is to recognize and avoid copying the accomplishments of other universes (the aesthetic adversaries we suspect are there in the multi-verse of other Bangs). The modest work I propose does appreciate the large distinctions between art and non-art—but it is also concerned with the historical strife—in this our world—between the overly praised and the unjustly forgotten.

My newly contrived blog might then say: "Let's shake the cobwebs out—and bring all the winners and the losers back inside! Let 's consign the old art-history to the fire, and make a new one—which regards all artworks as both special and equal." To be sure, this is an estimable notion—the social inclusiveness of art as a democratic value in a just society. Then, yes, we must include all artworks—which, taken together, project the sensibilities, aspirations, and sufferings of all the people. But there is this problem: Expanding the category of art so as to include all extant artworks, makes for an inventory that echoes the dire predictions of Malthus: It will, year by year, grow enough to eventually deny the very possibility of the categorical task that separates people from artworks. So what can we do, as people who want to live comfortably, to save ourselves from (being) art?

A Fourth Concern

No matter. I still have time to choose within the Collier clutter my memories have become, and there find contentment in the randomness and innocence of fickle joining that is unnoticed by the world. The work I now propose will have changeable parts, each mirroring a minute's or a year's obsession—one that can be discarded (but maybe not) to join the aggregate that is used to form a different work. I look for images that play below big presentations—those that manage to evade appreciation on the surface in order to swim in the deeper pools of acuity and belief. While such images, if they are unusually attractive, will surface on occasion to undermine the stability (by wagging their sleek tails) of sanctioned truths; they also, through their familiarity with change, suggest ways the public domain (by virtue of art's quicksilver prodding) can also be made to change—without capitulating and actually becoming art.

XIII

Ways of Making Art

Representation

For an interpretation that considers all things as representations, the sequence of art making that I describe here offers a material rather than a conceptual sense of limit. Instead of thrusting our perceptions into outer space, the artistic task is now limited to distinctions that can be made within a concrete format (a smallish canvas, or a semi-inflated balloon, or an unruly pile of rubble). This new modesty revives the perplexities of beginning out of nothing. But such efforts—whether publically or secretly ambitious, whether about the beginning of the world or the beginning of an artwork—must consider temporal divisions between the borders we posit as separating beginning, middle, and end. We can thus show ways of making images that describe how nothing becomes something, tarries awhile to make a world, and then fades into another (kind of) nothing.

The creative image is unremittingly in search of fulfillment: An empty canvas invites all possibilities of its despoliation or aggrandizement—but the verdict of satisfaction or dislike happens afterwards. Images of all kinds vie to show how they want the world to be. These descriptions are always partial—for who can know for sure that our present take on creation is the best and final one. But the ambition remains, and is carried on artistically because, despite its own subservience to time and style, art has a better chance of illuminating our later being—our longings to transcend the actual transition between the moments of our time—than do more prosaic accounts.

The past and future are not factors in animal life: "What might happen" is met by instant action to cope with the present of what is happening.

"Might-happen" is not an insight in the wild—although past experiences give animals habituated responses for dealing with what is happening now. But were such actions also based on recollection—memories of other encounters, and the ruminations they evoke—these influences would be a hindrance to coping with "what is happening now." This can be fatal in the wild.

Our own sense of what to do is given us through hopes and fears—which is an evolutionary advantage. We calculate so as to fix the threats of future outcomes through remedial actions in our present, as informed by conceptions of our past. But as we can only imagine such futures, our present actions take on their own rationales—skipping between preferred and discarded anecdotes of our past, and constructing manageable rather than chaotic (or ecstatic) images of our future.

Unlike with animals, the present is the most unruly state for humans—for we have learned that it vanishes the moment you acknowledge it—the present is faster at becoming past and transforming into future than our consciousness can retain it as the state of "here and now." When is now—and where is here? We would like the present to be most real—although neither in duration nor location does it stay put. We live in the present, edit the past, and invent the future. But our invented futures seek their corroborations in our past, in what we think we were and how it joins with what we want to be. In this scenario, the present is merely the facilitator—not the center—of our lives.

Our present has to be schematically, if not experientially, understood in its contribution to a theory of the middle that divides beginnings and ends. Boundaries are required—for beginnings often falter by misreading the road-sign of optimism, and so present the middle with a lower path—which leads to an unwanted end. Given the plethora of paths offered by our past, there is no good way, despite what the seers may say, to look around our present and find out where in fact we are—whether just starting, in the center of our middle, or coming to its end. And despite our best projections of our futures, chances are that none of these will come to pass.

Procedures

The scenario that is my subject: origin, middle, and end, can only be presented in conceptual, not chronological time. Everyone has a scenario for the living of a life, yet the emphasis on each part will vary—together with the time each takes. Birth—the expulsion from the garden—is the crowning moment. It is the time of first reality and the urge for change. Alternatively, the center of one's life—the passage from dependency to maturity—marks a more deliberate moment, valued especially by those who wish, if they could do well enough, that they could live forever. The end, in contrast, can best be measured by those whose life, for some reason (there is no "good reason") is already leaving them. These include the ones who simply die, the few who remember having lived a good life, and the many who wonder what, short of dying, could be as unfulfilling as was life.

Artists, in their creative moments, recapitulate the anxieties of this process: Origin entails the recognition of vocation and its attention to established style. The center is the transition to a way of one's own, where (in the light of possibility) one acquires the means of engaging with the public world. The end is judgmental—what one has done is compared to those who once were idols but who later became adversaries—the ancestors we first must kill (if we can) in order to fully die. The limits on artistic content are absent in the origins of a world, they are deeply evident in the center, and become equivocal towards its ending.

There are images in art that take us from nothing to something: Michelangelo's separation of the dark from the light, in the Sistine Ceiling, is the best I know. His creation of Adam—that touching of fingers—is about human beginnings, but not of anything that was there before. No paintings I know are about God's pre-universe dilemma, or about His Adam and Eve fulminations, or His take on how the universe is going. The down-slope into a final nothing, in contrast, is full of candidates: An art of ending means convincing us that less is more, and that we should pay more attention to the questions of "where," "what" and "why".

Stylistic imperatives (and recipes) for artistic "being-in-the-world" were plentiful in the early times of Modernism. Avoidance of their demands and suggestions branded one as a reactionary. Michelangelo might well have been ignored had he paid attention to the non-corporeal content of pre-creation. Later—during the first fragmenting of Modernism—the acceptance of a given demand became an indication of the parties to which you are invited. The word soon came out, however, that excess allegiance to a single style limits one's invitations to the many parties that now reward the new ambition. Flexibility in accommodating competing demands became a good protection against ending too soon.

I now present a work of mine. It is a hypothetical work-in-progress—which might be sufficient. In this work, "nothing" begins through an expansion of the made image to beyond the limits of the canvas (not only pictorially but physically). The image can then accelerate (as did the cosmos) out into the hall and down the dirty staircase—where it incorporates the remnants of last night's frivolities, and slides into the streets where it can rest while it basks in the morning sun—before the sweepers come. The parts that once considered themselves as autonomous, now crash the party—they explain that are caught in the new paradigm. As such they must, like the rest of us, merge with the traffic that these days is going all which ways—one tendril probing the perils of Las Vegas, another sliding a warm boat into the frigid waters off coastal Maine, a third staying put on Allen Street where the dreams of personal art replete with pubic hair still fill the early morning hours. There is no boundary to demarcate these works: One attraction is that nothing specifies their limit– no frame, no place, no pedestal—nothing one can offer that is not permeable in principle.

If the above scenario doesn't fit your tastes, I have a different one for you: Go back to your old infatuations with nature (kissing trees and talking earnestly to bears). Then imagine yourself walking where you once said you were not afraid to go. Protected by your memories in the early stretch, you eventually reach the end of the marked trail—where signs written in the local language indicate the places you might get to if you are willing to brave wild beasts and steep gradients. You have come to this point one time before, but then you chose the shortest and most level way that led you back, right back to the communal campground. This time however, you are older and fortified by an accumulated indulgence in the spirits of non-being, and so you decide to choose the hard way—the trail that has no name but a high degree of difficulty.

This traverse is important in that it shows your ability to endure and also to improvise when the trail moves across a ridge—that narrow icy path with snow-puff shoulders—before you reach what the sign says is distant resting place. But don't be fooled. You have come to a geographic equivocation that only leads to a more slippery and windy ridge, and then wanders, if you can get there, to a place you cannot see from where you now are. This furthest place (although the world is full of such places) is not really a place at all—it is higher and colder than any of its siblings. Reaching it is not to be attempted lightly.

It seems these days that for any place you want to reach, there are snow-banks which provide no footing on either side of the ridge you must traverse: Yes, yes; the danger is a devilish seductions for the novice—even if one hires a guide. But, thinking back on it, you were lucky. Do you remember that moment when your foot slipped, and, before you slid (with the guide's helpful yank on the rope) you fell on your stomach and regained the middle? Then, you tottered slowly—all bravura gone—to that lower place (the Swiss are practical) that is a train-stop and café for enervated tourists. The guide kindly says while he releases your belt, that this sometimes happens to the best of us.

But if you do get across the ridge—firmly treading dead center between the snow-puffs—you will become a chosen one among the brave souls who venture out beyond non-being. You will be a celebrant of the Big Bang in a Swiss lodge—a beginning denizen of our small world—one that has developed the capacity to move from free-floating particles (a slip) to directed consciousness (a save). You are also on the way (time being what it now is) to the end.

There are many beauties at the bar—this being the height of the rutting season—contesse, baronesse, marchesane, principesse. Your offspring will then be the last descendants to view the crossing between something and nothing. But these autumnal youngsters don't care. Having made much of inheriting the property of being human, they found that there are no unclimbed mountains left, and so, little else to learn. "Nothing more to learn" is a good metaphor for ending.

Studio Strategies

If I want to realize the tryptich I am planning, I must use the unskilled labor—mine and others— that is scattered throughout the dusty corners of the present art-world. I can't do it all myself—not as well as those immortal ones (who gave me the idea in the first place). Of course, instead of making a painting, I could mount a performance and call it "The Way To Go", replete with a bellicose old-master (me) and diligent assistants, (anyone I can find who will work for nothing)—and I would play it out—to the final glaze on an uplifted finger—even if, at the opening, no one shows up to even drink a glass of wine. But the old ways run deep; Tintoretto, Rubens, Picasso and I— we're stubborn. I could follow their arrogance, stretch my skills, and make three large paintings, one for each theme, and hope that the ambition of my subject makes up for any residue of awkwardness in depiction. (I still have trouble with foreshortening—but am getting better at kneecaps, and I have recently found that figures and trees sway much the same way). Of course, I could ride rough-shod over my ineptitudes, and with a lot of muttering and splashing, paint all three themes on one enormous tri-partite canvas—giving my few assistants the task of the drapery—leaving the figures, ah yes, for me.

But this is but a dream; it can't end well. (I know, I've had such dreams before). So I will try yet another way, one not burdened with the histrionics of pride and history. I will dismiss my pieties (and my concern with handiwork) as being utterly old-fashioned and sentimental, and dress my ideas (which I do not believe are sentimental) into a trans-media collage of historical sources. I am talking here about an art of everything—about extensive use of reproductions of old masters combined with snippets of my early work, together with writings of favorite philosophers—which I would precisely copy with a sumi-brush. Don't worry! I won't seek to desecrate actual masterpieces for inclusion—although, if I were a committed revolutionary, I would welcome a chance to storm the museums, despoil those images of beauty and power by cutting them into pieces—and then fit them into obscene collages which deny their earlier claims to excellence. But I love great art too much. And, anyway, such arrogance is also old hat—just drowsy winey sunset dreams tolerated by sundown friends. Reproductions will do just fine—they can be discolored or distorted, or made larger and brighter than their originals—and the cutting and placing is easy. But even so—even with access to electronic wonders that incorporate the best efforts of the last centuries as raw materials—can it be that my ambition to create an artwork embodying everything—will still be beyond my reach? Why, yes.

But there are limits on the other side that soothe such failures: Much of the world is scabrous, rotten, greedy and brutal—smelly, smug, and savage too. This, to be dealt with, would require an inventory of horrors that is more extensive than the one I could—or would want to—muster. I do not want to be old Goya—I cannot—my own past is still with me. But I do have time to choose within the Collier clutter my memories have become—and when I wake, I can prowl the back-alleys where contents lurk that are sure to fit inside the limits of my narrative.

My revised work then, would be in keeping with the unique gifts found in the detritus of different neighborhoods as filtered (of course) through my preferences. Such detritus comes separated by the usual divisions of wealth, race, ethnicity, religion, and diet. But they are all wondrous gifts, these cast offs and outs, for although they offer incompatible parts, each mirrors a specific obsession. Imagine what could happen when they all come together. If some are recalcitrant (still perfumed and pretentious), they can be sent back to the dumpster—or if transcendent (rank because unjustly discarded), I will make them centerfolds in my journey towards timelessness. For good material, I may have to range beyond the familiar alleys—go to the boondocks beyond the suburbs, and into tiny towns along the river. But then, having already been there, I can poke around without guilt (I'm just looking for my lost dog, officer) in my search for rejects that still show the obsessions of earlier times—but are now ready (sufficiently seasoned) to share their selves with me.

So what am I after? Why, the truth, of course! What else is there? Yes, yes—it is in art as it is in life. Even so, in both these realms, truth is best understood as a series of choices about what seems to fit—the hot and spicy porridge for our tastings of today. Because I am uneasy with historical absolutes, I will not go so far as to count aesthetic preferences (even mine) as encompassing what can truthfully be said or shown about the world. Art, sadly, cannot do that—although, at times, it has tried mightily. Instead, I look for images that play below the surface truths—those that flaunt their lovely tails above the water, but then swim slowly into the deeper pools of uncertainty. I follow them, those tasty slippery omens of improbability (but not impossibility) even at the risk of drowning in depths too deep for my understanding. Has a mermaid ever stripped off your mask to kiss you with her salty lips just before you drown? No? Then dive deeper—she's there, she's there, I swear.

I myself have swum a lot—with not too many kisses of the right kind for my efforts. I still try however. Here, I have here conjured up three themes that could explain it all. But before I go on to later stories—episodes of life as art that variously document my efforts—I must look at the themes in their simplicity once again. The themes endure, and indeed have grown larger through their descriptions. But I do not quite know whether this growth is benign—whether it shows insight into possibility, or is merely generated by fantasy. So now I must circle around my themes—a sort of Husserlian "epoche"—aimed at presenting everything at hand to a directed attention. This is like finding the art-world within the actual world. I have become wary now that both have grown so large—but their increasing girth is a warning to revert (from time to time) to reduction in order to find the limits—and thereby the specific content—of my themes.

The Beginning in Art

Beginning is a good subject for art. It provides the special reason for our existence; it introduces a first sensation that befits our feelings about the way our world began. Sight and sound can opt for early membership in the transition, for these sensations have the distance from us that is suitable for our experiencing objects outside of us—the world into which we are born. The heavenly host was privy—we were not—to the rehearsals of the lightning and thunder of creation. The earliest Gods were content to see and hear the beginning, but they, being plural, were only minimally aware of the ways things are. The use of the intimate senses—smell, touch and taste—arises later, when our world is more compacted—when we have to snort, suck, and shuffle just to live. So the Pagan Gods—just as we—had to wait for the world to become constructed with discernable dimensions—in order to satisfy the needs of knowing.

Time is an antagonist of the eternal heavens. It does not stay in place—some say it flows. The flow of time enables the development of early aggregates until they are complex enough to result in sentience—and, yet, that same flow, because we do not know when or why it started, prohibits us from knowing how it is that we are here. It also prevents us—because we do not yet understand the velleities of time (whether it starts and stops, or is only an illusion)—from knowing where we are headed. Strangely, it took many millions of years for the world to produce those entities (us) that are confronted with such dilemmas.

The Middle as Art

In contrast to beginnings, the middle overwhelms the senses, for it is everywhere—it covers the accumulations of our past, and is fertile ground for projections of our future. We deny—as we are inveterate self-protective middlers wallowing in the many-dimensional present—that we will someday reach the end. We do this through the stories we tell, which incessantly meddle with other narratives of our past, and we also do this by paying scant attention to the uncertainties that our present imposes on our future. This would be a fine approach if we didn't ever die—then we and ours would not need uncertainty. But as it is we are mortal, and mortality calls for justification—the how and why of our existence—which, in our hubris, we cast into celebrated images—until we tear them down to make room for other (we say better) ones. But all such images, while politically useful, are never to be completely believed in.

The most popular images of our time offer a novel immortality with every advance in media—and these offerings become more strident with every expanse in audiences. This is what the middle is about. The aim of advertising is to have us die thinking we are still alive.

The End in Art

We cannot luxuriate in the middle without worrying about the inexorability of ends. Just think about the fragility of life in your condo overlooking the Pacific Palisades: "But I have good insurance and the best doctors, and I demand an answer—I'm rich enough—why does my living have to stop?" No answer is forthcoming. But be comforted, dear client—ending presents a special case, for it has not happened yet—not even to you. In defense of our present living, we will find ever more subtle ways (just think of all these wonderful machines) to predict that it will happen later—oh, much later.

Given our conceptions of beginning and middle, predictions of how we end call for narratives that are constructed of both remembered and invented pasts. We also have a plethora of disputable presents to choose from in our need to assess the future. We must, we say, have some control over all this prolixity. We must find out the truth about this matter of coming and going!

The issues of origin and purpose are like parentheses around our present—but they give us different problems: As we do not know why we are here, we search our language to see if this is even a respectable question. Although we are ambivalent about optimism as regards our future, most of us are not willing to be penitents—to hunker down in the badlands, scrape off the moldering jollities of life, and turn—whether left or right or up or down—to peer at what remains of actuality.

XIV

Reasons for Making Art

It has been said that art, if it is not about truth is yet a precursor to sober explanation—that the changing forms of reality, and their modes of apprehension, first show up in the images of artworks. This should not be surprising, given that art—even in its most academic moments—seeks, in principle, to "go beyond" the style—and beliefs—that it inherits. Such goings may take centuries—as in Ancient Egypt—or it may take a week or two—as in our times. Changes in art seldom roil the political agenda—art's symbolic wars (pace Plato) do not much affect the polis. But radical art does provide good conversation both before and after the political facts—perhaps even justification for the use of such facts—given that artistic transformations have often supported the exercise of political power. But art may also compete with power by revealing different forms of belief—thereby inviting censorship or prohibition.

In its populist form—as advertisement or propaganda—art provides justification for ideas and their commodities, so contributing to their attractiveness through its dissemination. In this guise, art denies the finality and the loneliness of ends—you need something better than that worn-out sofa you now have. Getting something better—a move confirmed by millions—will certainly make you happy. And, unlike with life, you can always return it to the store.

To counter-balance the cheerful optimism of popular art, I turn to a different aspect of ends. Ends, in contrast to beginnings, need not be enamored of changes in belief; they can be tied to images of futility and emptiness—a metaphysical reaction to the fear of death and the waning of purpose: Most likely, we and our children and our children's children shall

all die; quite likely, our world will eventually become a cinder; possibly, we will all blow ourselves up beforehand. Such images support the turning of belief to the melodrama of last judgments in which our Star-Wars fascination with power and pain gains comfort in a projected afterlife of a cosmic heaven and hell—success enjoyed and failure suffered—a model of how it was on earth.

But as we are still here on earth, we can take the short view and downplay our fear of the future—the fear of "what happens after happening ends"—because we are still happening, and that (for some of us) may be all there is to it. If such disclaimers are not convincing, we can take on the artist's mantle, and envision the imbalance between our thinking and our ending as also a world-condition that can be rectified by images. We can say: The world ends when we end. We are then free to adopt the guise of tranquility that might, just might, help us peer into the voids of before and after, and find images that show how it was and will be. These efforts are secure in the absence of actual conditions—for we cannot know them. So, e.g., string theory, or multi-verses—as with depictions of the Resurrection—present images that need do no damage to truth—although we must be careful not to make them into bad art.

We begin our art-making with nothing; so we should end it with nothing's image. The emptiness of nothing is both inside and outside the canvas—not being anywhere helps it along in its eternal conflict with concreteness. Nothing, when inside the canvas, is a symbolic non-existent, nesting neutrally between the blank primed surface and the viewer. This surface, on a non-art reading, may be non-symbolic—merely a white cloth placed on the rack of a stretcher frame. It is your call to make it art. If you do go for art, rock back and forth upon your heels to find the precise point where symbol and surface reciprocate. You may not find it— it is like the rabbit and the duck—you can't have both at once. Yet you can find the image within the canvas—through directing your attention. You—not the laws of optics—have to decide whether you want a pristine symbol of nothing, or a tidy piece of cloth. The canvas, at this point of emptiness, gives you no options—and don't ask the medium for help. Neither glaze nor impasto, if you don't look at it as (for) art—will do the trick. Any piece of detritus, properly attended to, can do better.

There are many nothings. When emptiness occurs, some conditions of nothing can be found in the inertia of abandoned housing—or the persistent sameness of vacant faces—or the muzak defiling public places. But this too—although perversely—can be made into art. As I say: Just redirect your attention. When nothing avoids art by going into hiding, it seeks refuge from both inside and outside by adopting the guise of a concept of itself.

In this sense, nothing can be explained by its philosophical champions. But nothing can also be pictured. When hiding in the mind of art, it need not be unseen—like all else in art, there is always something that can be found to represent anything—even nothing.

The emptiness of nothing, if one brings it into art, can occur within the world of things that are traditionally limned by the picture plane—that encapsulated slice of surface upon which representations have long been made. Conversely, nothing can take on a modern undress, leave the picture plane, and have us follow it in order to distinguish its aesthetically charged non-existence from the ordinary kind. Any representation of nothing, however, raises questions about contradiction: Something (as artwork) and nothing (as content) can seem at loggerheads when pressed into an alliance. The result, seemingly, would be a full-blown contradiction—a violation of logical borders. Art, however, loves violation. Contradiction is the sibling of nothing. As such, it can itself be the representation which, when encapsulated and given flesh, is just what nothing needs for the alliance—the content—that allows it to be realized in (as) art.

The picture plane is less a surface than a codicil that seeks agreement between all parties—what is inside is art, and all the rest outside is life, until some art-ruffian breaks through the edge. Those backward looking artists who continue to look for nothing by making it into something within a canvas, also need to make a distinction between what is too little (vacant art) or too much (intrusive life). They, of course, have history on their side—with its ongoing pageant over the centuries of consecutive emptying out and stuffing in. Artists should know about this: Repetition without history can be dreary work. Who would want to recapitulate the nothing of unchanging verities—keep doing "the same old thing"—without knowing its ancestry? Change in style is not an enemy of "nothing-art"—it is an (inadvertant) enemy of its own dynamics. The flesh of life is mortified only to be supplanted; the slaughter of war is leavened by subsequent progress. Art changes too. These are dialectics that the school of nothing should consider.

For any who still wish to test their appetite for nullity by finding it in images within the sequestering limits of the picture plane, I offer some guidance. My proposals here are not rules, only observations—for rules are incapable of cohabiting with nothing. Yet, I feel that I must risk being (mis)taken for a rule-maker in order to protect my young would-be-nullist friends who, in addition to nothingness, desire to introduce some aspects (as above) of progress and success, and yes, glory too, into their work. So I couch my observations in the form that rules would take: I enumerate them, give them a declarative tone, and show how they may be inverted without their being broken. It is the most I can do before the ending that will come, comes.

Here, then, are my rule-like observations for artists who wish to tackle this issue of the future within the strategies of the past.

One

When seeking to paint nothing inside something, the developing image (both symbolic and physical) must keep its distance from the edges that surround it; only then can it begin to function as a metaphor for non-existence. The passage from heavy to light should fade before the material horizon of edge is reached. This does not signal the end of a journey, but rather a rejection of its boundries.

The journey's reach can then be reconceived—its end need not be the pit of individual oblivion—it's reach might even extend (images and concepts both have that power) to where the nothings, on both sides of your something, find friends in common with entities and endeavors that are indifferent to non-existence (say, an aberration in the canvas, or a bird feather that flew in through the window). This shows the fecundity of nothing— early reasons why it became something, and late reasons why it considers again becoming a nothing when it is still something

Two

To emphasize this independence of boundries, don't paint on the outer edges of the canvas—these, as above, are reserved for the first intimations of a metaphysically inspired passing. Of course, you may go too far—most artists do—but never fear. You can white-out the errant stroke that threatens the edge, and then you will experience the joy of erasing—of correcting your early excesses (by whiting—or, perhaps, blacking—them out). It is amazing how little one needs for an image when one starts erasing the leavings of one's early enthusiasms. But erasing has its own problem, namely, that one can go too far the other way and so, before the party is over, all your new-found friends will have gone. Then, there is nothing left on the canvas but your own fine sensibilities—discreet traces of a smudge or splash. If that happens, reverse course—pile it on—until you reach the point where the party starts again, and both you and the not-you are back in equilibrium.

Three

Make the first mark of being by touching a brush onto an empty place, preferably in the center. That will signal (a cosmos-shaking event) the beginning of something. You should have thick paint when you start—impasto suits a beginning of things—then work the paint as you move along until you make a fulsome shape—the saturated image that fits well around a middle—it can be a standing blob or a crouching nude—no matter. After that, you will have to move away and slowly, slowly (this is the hard part) thin the thick paint out as you retreat from the center point—until the fussing calms, the surface flattens, and with your medium spent you will reach the edge of (at least-at last) your world. Let the juices of your efforts peter out together with the paint. The point here is that the paint should just barely reach the edge in order for your efforts to reach their end. This way, if you are careful not to overreach, you can discern a something in your nothing—just as it was before the world began—and then you can begin the next round. "Relieve your edges" the old guru said, "if you want your conceptual space to extend beyond the picture-plane." Pollock and Turner and Guston did this well.

Four

There is another option, however: As the probity of images in art is always prior to the rules that support their making, a reverse procedure can also be considered: If you are ascetic to the degree that will enable you to dissociate "thick" and "thin" from their material properties, the image you make could instead move in the opposite direction—from the (thick) nothing of non-being into the (thin) something of being. You could argue (as have many) that non-being is thicker, more basic, than being. One argument, relying on Classic sources, goes as follows: There is the contrast, e.g., between the Platonic Forms and their particulars. The Forms are the symbolic center and the model for things. Yet, in themselves, Forms are more transparent (and so, more beautiful) than any physicality—although they are conceptually thicker. Forms are replete with all allowable variations of their particulars, but they do not contain the concrete substance of any particular thing. Even as the perusal of its formal properties establishes the identity of a candidate particular (say, a chair), Forms themselves, given their conceptual origins, do not exhibit their own particularity—for they have no conditions of use (therefore need no concrete differentiation.) The Form of a chair evidently cannot be sat upon, even though it "contains" the characteristics of all particular chairs. This largess can be understood as a conceptual thickness—a use appropriate for art—whereas physical thickness can be understood as conceptual thinness. Physical thinness, then, becomes conceptually thick, and so it can occupy the center of your schema.

However! If you do decide that thickness and thinness so reverse themselves, and your medium becomes thicker as it passes its outer edge, then you shouldn't be painting any more—for you have broken the edge of a framework—and have thus denied the authentic limits of your erstwhile symbol, namely, "pictorial art." You have also transgressed the autonomy of art as "Being" by dragging it into synonymy with the world's "Becoming." You have become the worm that chews through the carapace of non-being, and is then confronted with a world other than the one (with all your obsessive chewing) you have aspired to. You will then need to believe that, in this new and frameless world, less is more and the pursuit of nothing is the best. But "lessness" masquerading as "moreness" is also a kind of mortification—watching the wretched flesh go slowly soft—undecided about its proper state—as you castigate yourself for eating so much and not exercising. The achievement of nothingness (accept it) is flawed. It is the urge to eat the world and not have to fill the empty places—yours or the world's.

Although they each offend me, I defend both notions of thick and thin: You may take it from one who has scratched on many surfaces, piled them high and pared them flat— one who has often confused the middle with the edge, yet has seldom reached the end. But let's go past the personals and think more broadly: Are these issues—lessness and moreness—relevant to art these days? Can we still believe that "hand-made one-at-a-time," or "muck around the middle and clean-wipe the edges," or "you must cover over all of what's out there," are still the way to go? The kiddies with their smart-phones (and all else that the weekly Punjabs of Information produce) can make more art in an hour than you—what with your ambivalence about thick or thin (and your indecisions, anger too, about retention or expulsion) can produce in the slow-moving life you're still in. Also, those kiddies (now become adolescents) can send their wares to the furthest corners of the pre and post-art worlds—without insisting on categorical approval. 'Art'—that now debilitating word—sounds good in sermons and in jingles, but has few other uses left. Here is a dictum I offer: "Art is a schematization of changes in a nexus so as to transform that nexus into something other than it now is." Not so hard, that—only an affirmation of art's continuing prolixity in revealing the places where it has influence. Still too hard? Try this:" I just love what you do—the art, I mean—but my tastes are more upper central than downstairs coastal—but we should keep in touch; call me."

Five

There is yet another way to go: Your art—yes, I'm still talking to you—can be instantiated in a schema of world-wide extensions. You should start it within your own (not the canvas's) middle, and then contact all your friends, and the friends of your friends in every corner of the programmed world. You can show them (with whatever you use for transmission) the purposive incompleteness of your start in the new art. Then ask them to create responses (additions, erasures, etc.) which they send you in their own diagrammatic modes (which you may or may not be able to decipher). This accretion will constitute a complete (although ongoing) work—yours and theirs. Your job, as both artist and editor (it's not an easy overlap) is to give every participant a summary of what you have received from the others. Then they can add or erase as they think best—but all such changes can be argued against (by you) and, if need be, restricted—by you. We must protect each other (don't you think?) from both the corked-up and the incontinent.

Once you've taken this way, you cannot, even when speaking for the multitudes, define any stage of the work as final—for the value "final" must now be abandoned: Art grows like living things—like weeds or people. Also, you have no guarantee (nor should you have) that you will remain the clearing house—the Magister Ludi—of this enterprise. Your screen may well become silent, with the action going elsewhere. In consideration for past efforts, you may be sent a box of turds in plastic wrap to commemorate the end of your term—or maybe, more painfully, chocolate chip cookies in an artful container, to thank you for the hearts you've touched. But if you (bless your heart) should win another term, you will then have to spend hours answering questions sent you by surface mail (surface mail!)—questions about incontinence as art and what became of beauty. One benefit of such vulnerablity is that the furthering of my proposal requires salutory work—deep breathing, cautious but determined introspection, and self-imposed tranquility—so as to cope with the changes as they come in. In passing, you can reserve the right to return submissions with the admonition to better articulate the truly unsayable, and re-submit the barely thinkable.

But this direction has become too negative, too morose for cheerful creativity. So I will show you how arting can be when the juices flow. The submissions that I now describe are not meant to be actual. They are fictions about how it is in the times when no-one, then some-one, then no-one is there. In their various ways, then, they correspond to my subject of three themes .

Beginning

We gather wisps and vapor-trails, and group together the larger pebbles on a sandy beach. Within their pattern; we listen for changes in tone and density of the waves, and then we wander inland, across the dunes, to watch the trace of bubbling in the brook—an early portent of the cataclysm that may or may not come. A difficult passage for creation here, is the nature of the aperture through which the universe emerges. Is it like a cloaca or a vagina—like a geyser's spout or volcano's mouth? Or perhaps it changes—grows hair after it belches fire—to give us comfort by our remembering the entrance to the womb after our birth has passed. Beginnings begin in many ways; sometimes with a bang—as in Hiroshima; sometimes with a whimper—as in a bordello or a nursery. I prefer the notion of a whisper saying "I'm here," and I commend it to you out there preoccupied with the "why" of your own expulsion.

Middle

In hindsight, the middle is a train-wreck on a quiet country track that has had little history of such things. Back then it took time for word to spread. Other derailments, on faster tracks, have since happened and been forgotten. But this one became a fable for its time, preaching the news of its necessity to all the common-folk who would listen and be washed in the blood of the lamb. The later times, after this one time, have been spent in affirming it's necessity to others by codifying the preaching and then extended it, often through pillage and slaughter, to all those who would not listen. Some, increasingly many now, will deny such teaching, sometimes because of its conflict with reason, but more often because they have been sullied, after first belief, by its increasing divergence from high style. As the outcome is not yet resolved (we are now on the descent to a new synthesis of warrantless true belief) it gives some assurance that we are still part of the middle—and that the art we propose, like the universe, has room to expand. We tell stories—some true—about how far we have come, but we do not know how far we have to go. But some of us tell stories about our getting there.

End

Our coming to the end cannot be given the task of encompassing all that has passed—not even through an art-work. The past is now too inchoate as a span of time. This was once the job of the ambitious middle—but neither

Dante nor Milton could tell it all. The end is now celibate—as old folks mostly are. So it no longer counts—as did the middle—on accurate history, or creative autonomy, or priapic aggression. The art of the future is apt to be impersonal—no sublime heroes around to demarcate the changes in style, no great seers to adjudicate the contest between genius and pretender, no pedagogy profound enough to limn the past or fortell the future.

Disclaimer

All this is not quite true (although I wrote it). I should be (as above) more cheerful. Art will out, however vacuous or repugnant its envelope—and life will follow. Oh, there always will be stars—flitting above the seedy places and emitting sparks into the suburbs. But their value in the end may not be as a new portent—but only a winning of the creative competition for the best image of ending—how the burnout should show our creation, expansion, and demise to the disinterested spectator in eternity. But it might be otherwise. Who knows—we might stay longer than we say! Then, the lights of our passing star will become truly celestial—not simply afterglows from our late-late show. In that guise, our world will appear as a pathway to the unthinkable future, in which "ending" joins with "purpose" and "beginning"—as post-temporal markers of the cosmic order.

XV

A Return to Origins

I MUST, YET AGAIN, go back again—in order to go forward—not so much for your sake, but to assure myself that I am still writing about my subject—the themes I have proposed here as a way of interfacing cosmic process and art-creative process. This ambition, as I am well aware, is a philosophical stretch. Indeed, it may at best be a critique, an effort from the margins to compare philosophical concerns with other more speculative ones which are less dependent on logical analysis. Both concerns require imaginative flights which, in both cases, are susceptible to obsolescence.

Stylistic change is a given in art—as is the sharp but fickle delineator of greatness. It may also be that scientific theories, in their more scrupulous ways, reveal limitations to their own holistic ambitions—by identifying stubborn and persistent dichotomies in their subjects. The limitations of science I take as difficulties in the analysis of entities whose limits are not yet discernable for any comprehensive empirical procedure: mind and brain, consciousness and its neural counterpart, creativity and value, personal freedom and collective security, are some. The limitations of art show themselves in the persistent historical revisionisms that confound lasting distinctions between good and bad art as well as between art and non-art.

In this writing, I offer less an analysis than a polemic on beginnings, middle, and endings. through which I describe some of what remains important in the location of our historical lives—our perceptions of the past and future, and of the origins and end of our world—in a way that asks how and why they are concerns. I offer these conjectures as images of transition—on the one side between hypothetical developments in cosmic process, and on the other as ideological changes in art.

The sequence is a given: We begin, strive, and end. Each of these themes, in the context of religious belief, is divided into two—each theme identified through its own polarities: Creation transmutes between non-being and being; Annunciation embraces both celestial indifference and earthly love; Judgment contrasts contingent hopes with an eternal verdict. As each dyad contributes to the triadic form of the whole, we have six parts—which would remain three if each theme were internally united within a single mode of explanation. But we do not have a single mode—for having such would deny the pain, the pleasure, and the consequences of the transitions between the parts. We must therefore, for each theme, take what is apparent in its parts, and then elide them into the mystery of their transition.

Before the Beginning

First, let us imagine an emptiness which is not the same as non-attention. We may be the only ones in creation to pay attention to creation's move from emptiness into meaning—and we have been trying, ever since, to find out what that meaning means. Emptiness as an image can shows where a void can be coerced (in some sense) into accepting (in some other sense) the marks that lead to the strokes which could (in their combinations) give it substance. This is the way, I think, that creation happened, and as we are its issue, it is a reason for us to pay attention to the strategy of our own creation—for this is also the way of art. We begin with an attention that avoids aggression or flight by transcribing the experience into an artifact, and then into an image that attempts to control (make sense of) what has happened. The procedure is always the same: We go from the extension of mark into stroke and then the congealing of stroke into shape, and then the transforming of shape into image. We need not be too grandiose about this—for it is only a metaphor for what artists have historically done. Artworld values contain the presumption that making memorable images is the goal of making art. But images are slippery—they can be about something even when they are of nothing. As well—marks and strokes, despite their carefree existence, can transcend themselves and become properties of potential being, (images of actuality)—but then they face the stylistic question of how to get there. To be fair: Images of nothing—which occur in the material world that has opened to the conceptual need of creating a cosmos—deserve a special due. As above: Images are always "about" something. But that something, as shown in our recurring fear of nothing, may only be a blip in the course of creation. It is then tantamount to "being nothing."

Piero painted a solitary form of Christ—with his occasional friend, St. John, equally alone, trickling water on the forehead of God the Son so as to baptize Him. Piero was not a minimalist, but he painted an ascetic God standing in the early emptiness of a pictured western landscape—a silhouette of barraness, but also of divinity and its pointed purpose—the striving immortal soul against the non-comprehending diversity of nature. Piero painted his subject this way because he did not want it trivialized through the particulars of nature—so he made the landscape stark and simple, without fecundity or season.

Cezanne, who saw himself as part of nature, painted the forms of his simple subjects: still-life, landscape, his wife, because he found something in the sameness of their structure that avoided the sensual distinctions of codified appreciation. But he was also obsessed with the accomplishments of the past—certainly not for emulating—but to find a point of reconciliation between his acerbic view and the historical truths of imagery (however different the subjects and techniques) that are embodied in the constructions of old-master art.

During the Middle

Go to where the anguish about God's absence in lived life seeks resolution in the actions of a new and personal God. The absent father, following an old tradition, comes to earth in a lesser aspect of disguise—as an obedient angel—and impregnates a chosen human, thus welding the absolute and the contingent around a physical manifestation of both: the Living God. Shiva danced, Jove was lustful, Buddha both passive and cheerful, Yahweh distant yet admonitory—but God-the-Father's interest was in resolving schism—through a redirection that would elevate the new tribe of Christ-the-Son into a catholicism of diverse peoples, with sacramental assurances that the miracle of joining God and Human would be continuously renewed. Christ had to die to fulfill the miracle, but his mother Mary's role was of a conduit so that the "fruit of her womb" could effect the joining of two worlds—the old one of anxiety about a punitive emptiness, the other of a rejoicing within a new and personal certainty.

I have designated the Annunciation as the symbol of the middle—where, in my reading, the western world begins its historical sway. This placement is not a matter of interpretation or of documentation—there have been many such—but of belief. One can treat the matter historically, moving from the simple fable into the world-crises that translate belief into power, and thus propose a causal-historical linkage. One could also, counter-factually, ask: What would the world be like if Jesus had remained an eccentric preacher, troublesome enough (like so many others) to be hung on a cross and then summarily forgotten—and Mary just a dusty, grieving mother? But that one time of crucifixion was a historically exquisite coming together of old-testament orthodoxy, Greco-Roman authority, agnosticism, and the portents of a new kind of God—a human God, tied by triune identity to the eternal God, but the one who, through his mortality, suffers, dies, and is reborn. The implications of this change became powerful enough to transform the world.

This is the weight born by the image of the annunciation—not the simplicity of its myth, but rather the consequences of that myth's power for change. There is a host of well-pictured witnesses, beginning with the shepherds, followed by the three kings—and then, after the fact, the Mariology that, in the belligerent centuries that followed, had notables of all persuasions vying to be included in her portraits so as to be seen as in the good graces of the Virgin. These images can be folded, as fragments, into the historical account—but the real theme is elsewhere. It is in the tension between the Old God of distant and punitive austerity and the New God who offers the immediacy of divine love. This image is not between the something of present existence and the nothing on either side; it is rather a transition—perhaps the most important in our history—in answering the fundamental questions: about the "where" of the beginning, the "what-now" of the middle, and the "why" of the end.

These questions cannot be answered in rhetorical ways, but answers can follow on their transformation into images—for images offer the beauty of an answer presented as an occasion, and so ask for intensity rather than verifiability of belief. This is why Plato distrusted images, particularly compelling ones. He preferred the beauties of the actual beloved—for these, like concepts but unlike images, can always be held to strict account.

Powerful men preferred being represented in art-works with the Virgin—more so than with Christ or the Father God (e.g., Van Eyck's "The Madonna with Canon van der Paele"). Politics intersect with piety. Mary is the effective bridge between divine and human; She is also the symbol of beneficence and love—important virtues to be called upon and demonstrated when the powerful embark on adventures of conquest and control. Many female saints followed Mary; but their sainthood was achieved more by martyrdom than by rhetoric. The historical Mary remains a mystery. The gospels were written by men; no priests or popes were female—still aren't. Woman's power was relegated to love—an esteemed virtue. But Eve, with her storied deception, was always lurking behind the male fear of women. As it often happened, love without power could make a saint into a witch—whose immolation, in fire smoke and screams, would then be enjoyed by all who are too threatened (or ambitious) to love.

In passing, I offer some protests against the parable of the Annunciation (and its issue—the "Immaculate Conception"). Christian chastity, as derived from the immaculate conception, has given us insufferable problems: Pure (selfless, non-physical, spiritual) love as exhibited by e.g., abstinence, is profusely documented in text and imagery as a religious imperative, but it has been largely ignored as social action; it remains an ideal looking for an audience. In contrast, contraception, pre and non-marital sex, group sex, gay sex—all that spilling of seed upon the ground—are burgeoning practices despite layers of institutional opposition. As a result, the contrast between private practices and public denouncements has taken on the characteristics of farce—from which neither secular nor religious institutions are immune.

After the End

Ends are less mysterious subjects than are beginnings—but their aftermath remains obscure. We have documents of the middle to nourish us—nothing but speculation for beginnings, and only a few good predictions of how the end will happen. True, Michaelangelo and Blake did beginnings well, but they had the implicit God-figure—the unmoved mover—to instigate change and move the mists along. The end of things is more physical; it is not a question of possibility, as was the beginning, but, rather, a matter of material causality. However, "after-the-end" is another matter (if, indeed, it is a matter at all). Differently put: Here you are; the Old One gave you Being—and what did you do with it?

Milton, in "Paradise Lost," has God say of his creation: "Ingrate, he had of me all he could have; I made him just and right, sufficient to have stood, though free to fall."

Judgment is a favorite preoccupation in art. A compilation of Last-Judgments would make a marvelous exhibition. But after enjoying their beauty, we have to understand that punishment is also an aphrodesiac—especially when garbed in celestial clothes. The screams of sinners in their final torment have the quality of absolution for the rest of us: "You're not there yet, and don't worry, you won't be if you follow the established guidelines for avoiding damnation. But, even so, Moe and Joe and Tessie too, you could sneak in the back gate of Judgment Stadium and find a bleacher seat. What harm is there to watch the theatre of crime and punishment being played out—the drama, if you had been less circumspect, or more brave, or more talented—that might include you? Do you not wish—if only for a moment—that you are there—writhing in tandem with the major sinners of our world—for all eternity to see?"

One problem with the last judgment is in its conjunction with eternity. Granted, being sentenced to life without parole is pretty bad on earth—you have only death to look forward to. But those damned souls in Hell have no death to contemplate, they have only eternity filled with the yowls (of course deserved) of pain which no physical body could endure. So the pain of Hell, despite the graphic glee of its multiple depictions—when we separate its physical from its theoretical part—is empty. It can only be encountered through the pain of the soul—which may be an oxymoron if there is no soul to hurt—or it may be a twist accompanying bad faith—the absence, after death, of God's proximity.

In its theological aspect, the damned soul's pain is without redemption. But then, if access to heaven is forever barred to the damned—thus giving rise to our secular license to disbelieve—and the saved-dead are sequestered in their goodness, then why, in fairness, should the scenario of pain down below be allied with the pietistic bliss up above in forming a final mandate that does not change—that is defined as without time and without end? This "final separation" between good and evil seems to me too dualistic for Divinity's final word.

Note also the stress on vertical layering that these images provide—quite elitist in their stratification: Heaven above and Hell below. The fever-ridden, sex-obsessed, impoverished settlements down by the river, give way (if you can manage the steps) to the airy right-thinking sumptuous castles perched high upon the hills—those were the days. But in a less fanciful contrast: The antagonisms between secular ideologies carved through protest, and privileged states attained through anointment and lineage, are not always vertical—although they once were, and from time to time would still like to be. The powers of control are fast becoming horizontal—they are now between left and right—between differing interpretations of the world's historical development as seen in political battles about ways to destribute wealth, achieve fairness, and so on. Basically, these battles are between ways to hold fast to what we (think we) once were, or go beyond to where, at some time, we might (like to) be.

Why not then, dispense with all this inflammatory finger-pointing? Why don't you, God, let the sinners—instead of forever roasting in hell—come back to earth as laborers in our living fields, and have the blessed saved hold sessions to discuss what are the minimal benefits to give the damned—whether unbaptized or undocumented or sin-ridden—after the daily work of expiation is done? But, if I hear you, you say that the request is not reasonable; sinners are eternally guilty; it's a matter of free-will and its consequences. Some fail and some succeed—just as in the free-market: it's up to them.

Artists have had trouble picturing God because of this, less so with picturing Christ, little problem with Mary—and the Devil looks increasingly good on film these days. The Old God is still identified with Yah-weh, whose insistence on non-imagistic privacy gives Him little pictorial access to our imagination. But the Cross did become a potent image of the western world, and the interplay between Christ as son and Mary as mother—and the uneasy distance of the Father—are underlying themes, however disguised, of even our secular beliefs.

A Way to Go

Send out the word that the world, for its salvation, must become art! The best way for art to go is to recapitulate the processes through which the world began, what it has been doing since, and how it will be when it no longer is. We now have the Internet, which has no fear of eternity, and no stake in what becomes of us—or it. It is the paint-brush of our time; it is capable of endless addition, subtraction, recapitulation, transfiguration—as long as you don't ask it why it's doing what it's doing. But then, the whys we use for our metaphysical "before beginning" calculations, and "after ending" ruminations, are acceptable just because their evocation is of little use to our middle times—they have the non-pragmatic lightness of fiction. But we ought not despair of substituting usefuless with fiction—we could, as an ongoing game, promote the search for a a post-existential epistemology—one that dispenses with our thinking. The questions this game would ask may not need the Great-Calculator for their answers, but we would hope, nevertheless, to find the proper language for a clue to the answers in our middle time.

Succumbing to the "proper use of language" can, however, be constricting; it evokes the puritanism of logical analysis, where questions at the heart of ordinary living were brushed aside because of the improprieties and poesies they indulged in. I hold that philosophy requires fantasy and speculation to leaven—but not reject—the insularity of logic. I believe in an inclusive language, beyond the discursive, that stretches to encompass all our utterings, soundings, and lookings. To understand the levels and performances of language for different needs is, of course, not easy—but it is what science, politics, and art, are variously about. The ability (and desire) to modify putatively unerring world-descriptions through poesy and song—and so facilitate the adoption of new symbolic forms—may be the single most enduring characteristic of our thought.

Consider the internet: We (humans) have been feeding it information about ourselves to the extent that book-situated amalgams can no longer match what the internet knows about us. The Encyclopedia Brittanica, our last remaining hope for comprehensiveness in print, is now closing shop. But if you want to know something, just look it up on the internet—google it. An answer is always forthcoming. From where? Who cares. Is it the right answer? Who knows. Actually, there are many answers for every question asked, and this forces you to re-examine your question—and find out which answers fit what you thought you had asked. It may not matter after all. The answers, even those you didn't think you asked for, can be revealing for questions you might later think to ask.

The world as art is a series of answers. The questions are not nearly as important. They are usually querulous or pompous, and cannot describe—not as well as can answers—what the world is like. This is the power of art—old good art and unsettling new art. It generates answers for unasked questions, and (if given the chance) it can reshape the world. But we must be ruthless here. Most art, evidently, does not do this; it is complicit with the strategies of inertia. It chews on the eraser of old questions, and is satisfied with a place protected from the quicksilver changes that unforeseen answers insist upon. The distinction between art and non-art is little different from that between good and bad art. Both require a leap of faith for their crossing. Such leaping can be found in the answers we seek—and our hopes for these, in turn, are captive to the dread of our inquiry ending too soon, before the issue is played out. But unlike our death, the "death of art" is fun; and it need not happen (again).

The answers we attend to should show how to get rid of the notion of "grounds"—the terra firma of our thinking. Groundless, we are relieved of our boundaries and our dread of violating them. Take to the airwaves! Swing on the internet like a liberation spider—snatching up bits of bytes as they float by! Then, regurgitate it all (leaving only what is required for bare subsistence) and send the rest in care of "The World As Art"— a non-profit movement started by people just like you and me, who have found that flying is both an early and a late capacity in evolutionary terms, but that the recent fear thereof is merely a confusion between body and mind. Yes, we can fly—in our minds and the world's eye—as long as we don't behave like Icarus.

If you don't want to fly, then sit on your little plot of earth facing your too-small screen, (keep writing—soon you can afford a bigger set) and send your contributions to the younger generation of high fliers. This will help make the web into a world—an artworld to be sure—that has more art than it has people. It's not that we now have too little art around! Goodness! But most of it, unfortunately, is rooted to the ground—you can hang it on the wall, or stand it in the patio. If you're lucky and get a grant—you can paint it on the outside of a building, or anchor it (all ten tons of cor-ten steel) firmly in a shopping mall or village square.

But, friend—if we still are friends—we too-fat one-time flyers can still do otherwise: First, we must reject all the dreary stuff that first succeeds and then recedes, and consider making only that which can fly on its own—the effigy that shifts shape without being asked; the omen that appears and disappears in turn or tandem—the beauty offering a glimpse of its incomparable behind.

Artworks have lifespans as people do. Sometimes, like trees which fall where no one hears, they disappear within the nowhere where they fell. So gone, they become participants in the realm where their nonexistence is exemplary in fulfilling unused spaces. But in order to to exist, artworks must stay within the bounds that show their limitations—for after all, art is also a commodity. In the old-time world, the work and its uses enhanced and protected each other. In our new world, however, art is on the web—seldom noticed unless, perhaps unwittingly, you press its key— and artists are the spiders—flaunting as they emerge to feast on the entrapped passer-by. This, by some, is called appreciation—a giving up of the vital juices that give substance to art.

But wait a bit dear friend. Now that I have made you all be-rattled, climb down off the web—my spider has no taste for you—and let us see where, in theory, we have come to. We are not any longer in the world where one frequents places that are designated as appreciatoriums for art—museums, galleries, and gala openings. Instead, we have come to where we can look around for ways to separate mind and art from the hungry time-bound hulks we also are.

One difficult subject for an artist is a self-portrait. Rembrandt and Van-Gogh reached a point of truthfulness by looking through style to self—and thus showed their indifference to their own profession. Imagine an over-stuffed afficionado of the time looking at these efforts and saying "No, no, it's not anyone I would want to know." This is the danger, my friend, that tells us to leave town. Out there in the empty fields, we will find out how much faster we can fly when the antagonisms between existence and nonexistence are merely preferences—as between the eternity of Plato's forms and the ever-changing river of Heraclitus. The answer you seek—the choice between the two—as the Sophists say—is a matter of taste. Would you prefer, e.g., to be where nothing happens, where the demarcations between the talented and the dullards, rich and poor, the drivers and the driven, are not only clear but immutable? Or do you really want to fly?

So what else do I propose (now that I have proposed the truth)? I propose an art that has nothing to sell, is nowhere in particular, and has left its linear history—but one which is conformable to flying in tandem with those who are willing (yes—it's scary) to fly their separate ways—and yet like to keep in touch with the land from time to time. I dote on touching while flying—as long as flying, but not touching, remains a metaphor. One can sit in a cave or snuggle beneath the sheets, or stand erect while preening in the mid-day sun, swim in sweet and salty lagoons—and still fly. There are all sorts of ways to fly. The birds I wrote of earlier, are helpful here. They teach us that the degree of lift is inseparable from its content: Some of us have better pectorals than deltoids, bulging calves rather than fleshed-out thighs—a longer reach, more wind. This determines how well we fly. But if we just sit on that stubborn rock, the birds will not stay with us—they will leave before their ways are made clear enough for us to copy. Without the birds, we must do it on our own.

My proposal contrasts with the scenario of straight purpose that would have us (despite the casualties) march inerrantly from beginning to end. Consider these other ways: Flying in circles (as the Buddha has it) or in spirals (as the avant-guarde would want) or any way the wind blows (as some of us now try to do) are answers to counter the threat that the world will end too soon—before you, my friend, and I have had our fill of flying. You are older now, and, yes, a bit braver. So begin your journey at the intersection where once you brought your sometime loves.

Other Ways

All this, however, is getting too oblique—too much like looking at oneself in someone else's mirror. So we had best change the direction of our thought. To start again, let me start with you: Imagine walking somewhere, where once when you were younger you had been afraid to walk. Yes, yes, I know—it's late now, the bugs are biting and it's going to rain—but this shouldn't make a difference to your larger seeking. This is the way (the new way we've always been talking about) that brings you to mountains replete with valleys that are only logically there, and to cliffs that persist while facing down the roaring ocean. But then, those battered cliffs and shallow valleys—tired after all these years of coping—will slowly back away and fill, leaving meadows that have stayed green on salty spray.

But listen! There are animals around! Growls and sneaky glimpses in the bushes show they are not pets—but they remain too far away for you to know if they will just preen or eat you up. A suggestion: Kill your campfire. Then, if you squat and shit and chant and wave, they will come closer without aggression, and show you what and how they are. They are large and lithe, too fast for you; but they are interested in your smells, and, if all blandishments fail, in how you taste—it's really up to you to pay them the attention they deserve. But watch out! There are others out there—hard black shapes behind the shadows, uninterested in anything but eating foliage—don't get too close; they are huge and horned and easily annoyed. The ground they tread is spongy in places, but they have wide feet to stay above the muck, while you might sink because you are more human. The high, dry, and narrow path is better—the horned-ones do not go there—although the tall bordering grasses surely harbor insects whose feasting on your bare legs will result tonight in fearful itching. (Why did you wear shorts?) The sky—if you are still brave and will look up—is a comforting blue, crossed with jet-trails from the ordinary world. But look again! Don't you see a distant line of clouds that is quickly darking. A storm is coming. Oh, you say, merely a summer shower that is good for greening in the valley. Or perhaps, (one never knows) it is the coming of the tempest that precedes the deluge resulting in the cataclysm that signals the end of the world. It is too soon to tell, so keep on walking until you reach the trailhead.

My good friend—in any case—don't fear too much. Just blame your anxieties (despite your posture, you were never truly that adventurous) on the fact that you cannot know whether Tigris will have his fill of you, or whether the lightning will crisp you into post-philosophic bacon, or whether, safely down from the mountain, you will celebrate by dying of over-eating in the valley.

But how, with such timidities, do you still insist that you can offer an unbiased account of big-bang action? You live in the world where we all, alive and dead, have also lived. And, recently, we've been taught that the Big Bang Progeny (such as we) have progressively evolved: We came (so this story goes) from the scatter of mere stuff—to sentience and then to rationality—and because of this to world dominance—albeit later into unhappy consciousness—more recently to mutual destruction—and to what it is we now are—afraid of the coming dark. In this ordered scheme, "before beginnings" makes little sense—no one was there; and "after endings" can only be met with political fatigue—as none of us will be there.

Now, having gotten this far into despair, you must wonder, my once optimistic friend, why you ever believed in the step-by-step scenario of innocent birth to meaningful death. You might even wonder about those pious others who would denigrate your last free act—your dying—by insinuating it into a programmed afterlife of Hellish suffering or Heavenly bliss.

On the other hand, it could be that the persistence of this fable is an attempt, by the more progressive gods, to hold our childish fears to ridicule, and thereby offer a better way to push us up the evolutionary ladder. The last stretch is clear, they say: Pass through the forests of your once-prized consciousness into a selfless existence on the post-rational plain—upon which are scattered modems of calculation that are finer, more rapid—and more beautiful—than even your finely arched brow can match. Then you will relax (as tigers eventually do in zoos) and concentrate on eating and loving.

Imagine! All this time, I have been sitting on a rock and talking to myself. I am still on an unfamiliar path, deep in a primeval forest with predators circling. I walk on soggy ground, slap at bloodthirsty insects—while suffering from a fragmented psyche, an opaque future, and the storm that is surely coming. Yes, I can sense the lightning and the thunder that announce the deluge that will wash us all away. Yet my fears can be mollified, and my journey redirected, if I adopt a different strategy: I will (I swear) forgo my concern with images of a judgmental nature—I will come to love the whole thing: predators and bottom feeders, sublimation and sordid sex, hairy warts and too-soon sagging breasts, dumb critics and flaky artists, corked-up priests and virulent philosophers. These are images that distressed me in my early living, and drove me to the narcissistic remedies of climbing cold mountains, looking for girls in seedy bars, and swimming too far out to sea. As a penance for having once considered such things as worthy of life-as-living (although then they seemed just fine) I now, in later life, plan to transform what is left of me into a work of art.

XVI

Savanna

THERE IS AN ARTIST you should know about. She was born in the deepest jungles of the Amazon. Her name is Savanna, and her tribal name is Iriguri. Her mother was a native girl and her father an Anglican missionary—who stayed awhile but then left to preach in safer places. Savanna's childhood was spent in the jungle, among the stinging plants and snakes and jaguars who also lived there; they were her companions none the less, and she was taught—more by her mother than her father—that these creatures are not aggressive for no reason, as sometimes humans are—but still, she needed to learn the balance between her needs and theirs so that they all could live, if warily, side by side. Because she was a missionary's child, she was sent to a school, a missionary school many miles away—too far for her to return, except occasionally, to her mother and her jungle home. She did not much miss her father. The separation from her mother was trying at first, but her teachers all believed that the salvation of indigenous tribes comes through a joining of civilizing life-styles and imparting true religious beliefs, and so they assured Savanna that the separation was in everyone's best interests—including God's. Savanna's classes relied heavily on picture-books and films from which she learned the rudiments of Western culture and the English language, and what it means to be ambitious. She was especially taken by the photo-displays of fashion and home-décor in her art-classes, through which her talent for pattern and color, first kindled by the native arts of her jungle home, found a direction for her later ambitions.

Savanna was a stand-out in her class; she was bright—although not overly verbal—and by Western standards, she had startling good-looks. She was happy with herself, aware of her gifts, and unimpressed by her fellow students. The girls were only interested in praying and cooking; the boys were all shorter than she, rather coarse and rowdy, and unswervingly priapic. Savanna remained a virgin all through her school years. Her attributes, and the admiration and secret envy of her teachers, eventually won her a travelling grant to America—which she happily accepted. Before she left, she returned to her jungle home to see her mother—who, curiously, seemed indifferent to her daughter's good fortune. She said only "Be careful with your money, watch out for talkative men, and come see me later—if you can."

When Savanna got to America she made her way to New York City where, through pre-arranged contacts and some fortuitous exposure, her looks and talents were soon noted. She became a fashion-model—particularly prized for her dark tan color, her feral yet distant mein, strong and supple back, sinewy legs and delicate slender feet. Her feet had previously been a problem for her. The villagers back home had flat sturdy feet with wide-spread toes; they could slog through the worst jungle terrain barefoot. Savanna's feet were long and narrow—a deformity, her mother said, which came from her father. In turn, her father liked her feet; he also liked to fondle them, and he insisted that she wear shoes—even though it limited her ability to do chores. Perhaps that was a reason for her mother's willingness to let her go; it might also have been a reason for her father's leaving when she was still young.

Savanna soon became a fixture on the runway, and was often asked to model, sometimes to pose and play, in Europe. As she did understand money—a trait that came from her mother—she would go on these junkets only if the pay was right. At one festive exhibition in Paris, she met a Stanislaw Albert, an international financier—mostly French and Polish—who was smitten by her many talents, and despite having a current wife, suggested to Savanna that they have a relationship. Stanislaw was older than Savanna—by some twenty years, but she was approaching the age-limits of the type of modelling she did so well. So, after some discussion about practical matters, she accepted his offer. Savanna was not sentimental about marriage, but she knew its legal strength. On alternate evenings, usually after champagne and an elegant meal in a fancy bistro, she would return with Stanislaw to the studio, and would quietly but directly describe the benefits of their marrying. Stanislaw was not naïve—he had after all become successful in a cut-throat business—and he had had many affairs. But he was acutely, viscerally, aware of the admiring looks, the leers even, by both men and women, that were constantly directed at Savanna. Marriage would not stop such gestures, as he well knew, but it would erect a fence, patrolled by legal Rottweilers, that would give the more lascivious ones some pause. Although Stanislaw was getting old, his sex life with Savanna became better with each passing day. He attributed this to their mutual love of kissing: Sometimes they would begin with the mouth and journey to the knees; other times they would start at the feet and end up well above the ears. Either way, it was always slow going with many stop-overs and detours. Savanna had usually been aggressive while making casual love, but she found the slow and ruminative way with Stanslaw more to her taste. It was like a filling in and emptying out—a merging—with the street noises as accompaniment.

Noting her love of pattern and design, Stanislaw procured for Savanna, as an early romantic gesture, a studio in Soho, which she furnished sparingly, used frequently, and began seriously to make things other than what they were. Savanna had no interest in such traditional designations and formats as painting or sculpture (or, for that matter, as between art and non-art). The notion of a frame or pedestal were irritating because it signified a separation between work and non-work—between what is made, or found. The very notion of "art" was even more distasteful. What bullshit is this? Everything in her studio was art. The floors, walls and ceiling were painted on, broken up, reassembled—no doorknob or chair was spared a gob of color. She had not as yet extended her interests into the halls and the building's exterior—yes, and into the streets—but she held back because of pleas Stanislaw made concerning property rights and lawsuits. Savanna often dressed and decorated herself in novel ways—taking care to avoid toxic thinners and self-mutilation—no tatoos or scars. She had now become overtly sexual, and she happily anointed herself with the best perfumes. Smelling curiously good was part of the art-form she was beginning to discover. She loved to paint her body—a way of separating it from her past—but it was also a way for her to physically be part of her creations. Stanislaw was first aghast, then bemused, and soon entranced by her changing appearances and her hot ideas.

As he looked at Savanna's accumulating debris, Stanislaw remembered the art collected by his previous wife. It was all about the elegance of nothingness. The lady particularly prized the recognition of her taste by her cognoscenti friends. After some embarrassing missteps—she had once subsidized "happenings"—she focussed upon artworks that were minimal in body but had extravagant claims to universality and finality. She lined the walls of their condominium with assorted rectangles in austere frames whose blankness would be modified by a simple smudge, or a discreetly painted circle, or perhaps a word—like "merde" or "fuck"—tucked unobtrusively into a corner of its monochrome canvas. None of her guests looked at them more than once, but they paid due homage to these works, and Stanislaw went along, using vaguely remembered existential phrases to keep guests and conversation from moving too quickly to the hors-d'oeuvres. But he did not particularly like these paintings. He wasn't sure; he knew nothing about art; but even so they seemed empty—without any depth of meaning he could see—also, they were terribly expensive.

Stanislaw never did like his wife. He had met her, years ago, on a skiing vacation in Switzerland. He skied poorly, but she encouraged him, having quietly discovered that this clumsy toad had money. It took some sessions in hot-tubs overlooking the Alps and dinners in the best restaurants in Geneva—where she knew the waiters well—to get him to ask her to return with him to America. "Oh no," she tearfully cried; it would ruin my reputation. I want to very much, really, but I couldn't—all my friends—unless that is, we became man and wife." So they got married, somewhere in a ski-lodge in Zermatt, and then flew to New York. Stanislaw knew that he had been taken—but at the time he didn't really mind—he was wealthy and bored, and the hot-tubs had been lots of fun.

The decision to divorce his wife and marry Savanna did not come easily. There were, of course, the money matters and the social natterings—and there also were the children. But the impetus to get away came from feelings that he had not previously acknowledged. As his wife no longer skied (American resorts are so vulgar) and was not inclined towards health-clubs (all that stench and sweaty hair) the visual aspect of their sex-life, once important to him, had diminished—and he did know about her many boy-friends.

There was something else: Stanislaw had always enjoyed a variety of erotic passages—he liked to smell and taste and listen—and above all, look. He resisted the beat of direct love-making, but was drawn to its various detours—the actions that slowly return to the (for him overrated) climax. Stanislaw actually preferred to be in many places—to feel a bit—just enough for now, and yet be free to watch the others feel more fully as they went along—so that he could enjoy the noisy abandon of the enjoyment he had given them. He did not think this a weakness—actually, he admired himself for the distance he could achieve without losing control of his separateness even in the most demanding of intimacies—a talent that he also used to good effect in his business ventures. Above all, he did not want to fall into a funnel where his center would be squeezed into a place not of his choosing. Savanna didn't do such things.

Stanislaw's ex-wife, now sitting in their penthouse in Manhattan, had shown little interest in satisfying his, to her mind, increasingly perverse and unfashionable needs. Also, she had early been converted—by her many admirers— to the "wham bam thank-you m'am" style of love in the American movies. Stanislaw, little fellow, had no way to cope with this. So, rather than suffer more furtive years of watching starving artists and itinerant cowboys mount her staircase, he decided to split—to leave that witch—and find solace in the sinewy arms and conical breasts of tall dark Savanna.

Savanna did not simply satisfy the fantasies that Stanislaw wanted for love; she made them her own—however many times they changed. She could grovel, she could coo, scream, curse, brusquely push him off, sing while coming, tell stories during foreplay, squeeze his balls until he howled—none of which were services, they were how it was for them to live together. She was also faster than he. Savanna did not need the "center" that New Yorkers talk about—indeed, she did not know the word until she had to comfort a few hysterics in her early beds. Her own childhood had been spent dodging snakes, killing the smaller animals who tried to eat her food, and avoiding the predators from the jungles and the neighboring tribes. She went through the activities that filled her days without a thought that she should try to be someone other in each one—she was herself in all of them. So when Stanislaw asked her—for the sake of his own needs—that she should appear to him in many ways, she readily complied—as she had become interested in the contrasts he was increasingly able to articulate. They were like those that she was making in her art.

The wedding ceremony went beyond what Stanislaw expected—Savanna had taken charge of the planning: The service itself was conducted by two priests—a black Rastafarian and a white Episcopal lesbian. The hired hall became a temple of symbols and artifacts from many cultures—hand-hewn benches, Berber rugs, throws and curtains of the finest filagree, large patterned cushions for the more adventurous recliners, murky baskets for unmentionables. And the tall grasses beyond the pool had biting bugs to please the penitents. The music was full of chanting, drumming, the sounds of jazz and rock, and some primal screams—all greeted by ululations of approval, much stomping, and uncertain pleas for help from the newly fallen. There also was a skinny violinist from the Philharmonic who played unaccompanied Bach sonatas throughout the evening. (He did swerve a few times to play Paganini caprices—most everyone noticed the difference).

The food was best suited for hand-consumption—but an assortment of sharp knives and two-tined forks offered a "back-home" image to the hunters. Spread out on large wooden tables were exotic delicacies, some still wiggling, from all parts of the world, sauces sweet enough to anoint your partner with, others hot enough to stun the live eel you were determined to swallow whole. On other tables there were large roasts of smoked and marinated pig, and aged beef fillets just barely scorched; also refried beans, wild rice and, yes, hash-brown potatoes sided by young arugula. The wines (hard stuff was not served) came from Brazil and Chile—a gesture to Savanna's origins, and a boost for a growing enterprise—but also, Stanislaw opened his stock of premier cru's. On discrete stands around the hall were offerings of kif, peyote, and the buds of fine cannabis, imported directly from Amsterdam—but nothing harder—the police had been assured.

The guests were chosen for their talent, for their willingness to talk about things of which they often knew nothing– and to make good jokes. As one would expect, there were many beauties, some eccentric professors, rough and smooth trade, the poor and the rich, the obsessed and the truly gifted. But none of the guests—and this was the red line of demarcation—was boring. Some were outrageous—kinky, often crazy, many were rich, others obscurely unemployed—but boring, no.

Savanna dressed Stanislaw for the occasion. She insisted that he not get a haircut for a month before, and not shave for two weeks. She dressed him in a shirt with Byron sleeves, a pair of fitted dungarees, huaraches, a multi-colored sash to cover his pot, and a large necklace of oyster-shells and semi-precious stones. With all this, he looked younger, slimmer, and, indeed, almost handsome. The realization of this change—his first-time happiness—stayed with him for years.

Savanna also chose a simple way: She wore a fitted gown of crocheted linen whose weave, as she specified, was open enough (she wore no underclothes) to show the undulations of her body—which she was happy to exhibit even as she dashed around to see that all was going well at the celebration of her first and only marriage. The night passed into morning and everyone was gone by late afternoon of the following day. Stanislaw and Savanna were now free. They were married and no longer had to face the travails of their previous lives—he the anxiety of acquisitions that would enhance his status as a financier, she the need to sustain herself against the blandishments of new-world predators. Her new name is now Savanna Iriguri-Albert. She liked the hyphen.

They both continued to change—he wanted to deconstruct the trappings of his earlier success, secure his various moneys into a trust, and begin to reach into the guarded places, beyond the allure of fragrant seductions, that he had already found in his new wife. Savanna also needed other ways to think about herself. Now that she was mostly free of the daily look and leer, she could spend more time in the studio that Stanislaw had given her. She could move from the passivity and numbness of modelling to the unpredictable adventures of art and self-criticism.

For some time Savanna had been uncomfortable with what passed for art in her new country: Unlike the celebrations in her jungle home, the gallery openings were predictable in their claims to novelty, but yet were embedded in the tastes of the newly rich (and the failures in taste of the older rich). The art-market was not (as she had once thought) an open bazaar—it was a salon in which both works and patrons could get facials and pedicures in tandem, and leave lighter but more connected (to something they had been told was there) than before they came.

Savanna did like some of the so-called outsider art, and wondered how the artists could be so free to repeat themselves, write on their work, and paint those awful genitalia—and lovely flowers—and obsessive patterns—while they were also certifiably insane. It was a new problem for her: Does one have to be mad to make great art, or only some great art, or only nowadays? Savanna had also gone to see great masters in the great museums—mostly at private openings in advance of general admission, where she had earlier been taken for lessons in good taste by her then patrons. But Savanna looked longer at the art than did the others, and she did come back (alone) days later, to look still longer. She saw that that even in the old stuff, art could be about important things—and she began to read a lot. She came to realize that the old and dead artists she admired were not all insane—they were obsessed with the possibility that through their skills, their art might reveal some trappings of important truths. She particularly liked the Northerners—Durer, Cranach, Holbein, Breugel and Bosch—for their familiarity with pain, their crabbed elegance, and their distance from the Greek-inspired images of the Italians. She didn't much like Michelangelo or DaVinci—being herself somewhat homophobic. But the great displays of the flesh in Titian's goddesses, and the histrionics of love in Tintoretto, were moving in their certitude that sex, when sublime, was good. Savanna was a bit jealous— nothing had yet challenged her capacities that way.

That stuff she saw was really, really good, but it is now time, she thought, to go somewhere else—find another way to separate art from connoisseurship and fashion, and regain at least the possibility of an art that probes the unspeakable and ecstatic—both the belief and the celebration. But how? Does not art always moved in its assigned way—the great works being spotted by the cognoscenti, and then put into the historical winnowing machine—from there to be distributed for the delectation of future cognoscenti? The results of this process, while not perfect, have so far been self-corrective within the notion of "great art". But there was once, to be sure, much art- not then appreciated for "greatness"—a modern term—as much as for the skill to celebrate extant beliefs. This was confusing to Savanna—the strange amalgam between the historical need for social affirmation, and the contemporary taste that has a different need—vaguely called "appreciation-of-the-significant-new".

Savanna often sat together with Stanislaw—she quite naked except for sandals, he in the Byron shirt she had made for him—and together they mulled the problem of art's present exclusivity and purposelessness. If, as Savanna said, everyone makes the art—even if only a small part of it—as in my tribe—then we are all artists, and the artwork need never finish because it always walks away to where the impetus to add is greater than in the earlier place. If someone there would ask "where is the art?" you could wave your hand and show them all there is that they can see. "More is coming" you might say, "but it may be going to another place—follow the river or the upland trails, but you can also look at the internet. The art-work goes wherever it can be added to—to make it better (larger, faster, less ordinary) than it now is. You are free to follow this course of art—although I can't guarantee your safety."

Stanislaw thought (fleetingly) that if he had had someone like Savanna in his early enterprises, he would be worth billions rather than the paltry two hundred fifty million he now can draw upon. No matter, she may take off her sandals soon. Later, when she had wiped him off her breasts, Savanna said: "We are coming round to where the divine Architect tells the earthly architect (I learned the difference in missionary school) to bring together painters, sculptors, poets and musicians to create rivers of art that will flood the world and first bring life and joy—although later, in their downwood turn—when they become dry—they bring sadness and the fear of death. But art, despite what they say, is always catching up with life—although the living seldom acknowledge this. Rivers flow past your wading place—as some old Greek said—so don't expect, if you are fishing, to net masterpieces that you can dry and hang upon your wall. It's all in flux, art and life and oceans and fish, and should be so—just because the rapacious ones will have to change their ways, or do without the fishes that are flowing by. The best fish, if you can catch them, will one-day be worth lots of money—but the repercussions of your misses will hurt you more: "What? You didn't notice the Van-Gogh for sale at the Happy Cancer benefit? Just because it looked so crude? Well! It's up at Christie's now—although earlier you could have snapped it up for much less."

Stanislaw, beloved husband, we should go with a different flow—we should live art in the way we watch driftwood on the Mississippi—floating slowly past while seeking to avoid aggressive boat-propellers—until the river floods, and then you will find twisted wood, snakes too—all art—in your own back-yard. So let us write a manifesto (I have already made a rough) and put down some rule-like observations that are meant to be obeyed, added to, and also broken.

First: The "art-work" is all around; it has no place and no time—except for the places it inhabits (most art-works rent) and the times when you pay it the attention that defines it. Art-history is a concatenation of attentions paid—within a place and time—to entities that seek to and succeed in becoming art.

Second: Art has no form or content except for that supplied by a specific attention. So don't try to hold it fast, tie it up or down, protect, restore, or suffocate it with praise, blame, or analysis. Let art become non-art—as in democratic societies—or, as with work done under totalitarian regimes—bad but useful art. But both these judgments (the "non" and the "bad") will change as time goes on. Re-appreciation is an underrated aspect of art history.

Third: Art has no intrinsic value. You can bid it up or sell it off. You can give it away or throw it out. If you are uninterested you need not notice it at all. But you can also yawn and pass it by. Defenders of museums and other fragrant mauseleums say that their collections of art-works are the guardians of human history—that they are records of sensibity and social aspiration. This is a truism. Art-works are such by definition: Some expand a way the world can or should be known and some do not—and there are so many, a clamour of them, now.

Fourth: You may say of a candidate art-work: "There's something to that—or I think that's beautiful." But for there to be a "that" to which the "something" applies, is to have you pay attention to it. It may then become art—it's up to you. Think of it: You can be an art-maker without actually making art! But don't look to the aestheticians or to other artists for corroboration. Especially, avoid the dealer and the critic.

Fifth: If you say instead: "There's nothing to that," it does not imply an attribution of non-existence—or even non-importance. It is rather a criticism of the "that" (poor thing) because the "something" that had previously been applied to it is inaccurate or un-deserved. Your attention has been misplaced—you have been snookered. But then you must decide whether the recipient of your "nothing" (the "that") is non-art or just bad art. In either case you can round-up the artist and the dealer and the critic and bawl them out for not telling you. But also, and importantly, you must clean up your own criteria for paying attention.

I didn't make this up all by myself, dear Stanislaw. I learned some of it from that wise old recluse, now living down by the Mississippi River. He writes books and makes paintings too. I wanted to learn more from him—thinking he was only interested in talking and watching the river flow. But he did look at me from time to time—that way, you know—so I left.

XVII

Edvard and Umma

THIS IS A RETELLING of an old story about a missionary who went to the far north—the Eskimo regions—to save souls. I had read the story years ago, but I never did find out who wrote it. I was too young to look it up—but it stayed with me in drifts of snow and dreams of blubber and an Eskimo girl who, would you believe it, overshadowed all the girls I knew in Brooklyn. So, because of love's persistence-in-the-mind, I now feel free to tell my version of how it happened for the others. But this story is also a sequel (or is it a preamble—I don't know which is right for now) to a story that I have been trying to tell you all this while—between each of the times I was able to step between the admonitions and evasions, to expose the crooked straight for what it is. I also don't know if the story I tell you here begins or ends in tandem with the others. But in its fashion, it continues the themes I have here been working on. Stories are sometimes better than arguments in this regard. By telling this story, I want to offer some other images that can flesh out the ambitions of my larger theme.

The missionary of my story travelled to this cold and distant place to save souls. He was not practiced in this duty, but had ventured it in warmer regions where the sun and the styles of undress made the spotting of souls in danger an easy venture. His present mission (to way the hell up north, where everyone is always overdressed) was a mandate given him by his superiors—which, by fiat, had to be obeyed. The mission, they told him, was in response to the region's needs—the fish were dying, the seals were leaving, and disbelief was spreading through the villages—a good climate just now for saving souls. They were tiny, these villages, and getting smaller every day. But the salvation of just one soul, for his superiors, was a sufficient affirmartion of true belief. Our missionary, according to self-examination, had saved many souls—but there were some he had admittedly misjudged—they had led him to look deeply at their own (vile) forms of salvation. So he dutifully (and with some concern for the salvation of his own soul) left the warmer regions and took up the call for a mission in the most primitive place that his faith had marked as needing God's attention.

The missionary's name was Edvard Gurisiewicz, born in Brooklyn, first generation, and captured by the church before he had reached seventeen. He had little memory of his years of indoctrination—but they were better (more sequential) than the wide-open streets that had oppressed his childhood. Edvard was called Eddie Guru by his friends in the monastery (because of his interest in exotic beliefs) and Father Ed by the parishioners he tended to. The locals liked him because of his overt faith and ceaseless energy, but they were also somewhat relieved when he took on the missionary's role—because, in truth, he was an unsettling presence everywhere he went—exhorting and demanding and cajoling in ways no-one could quite understand—he talked a lot to women, and some even came to think he was quite mad.

When Father Ed arrived at Shishmaref in far Alaska, he had to take a smaller boat to a village even further north, whose name he could not pronounce. Once there, he was greeted with hospitality by the chief of the region, given seasoned blubber and seaweed to eat, and was introduced to the villagers—who were happy that the new missionary was a stalwart, capable-looking man (not like the other one, the skinny-scrawny who was always trying to bless young boys). On the third night, after dinner, the chief invited Father Ed to cohabit (as was the custom) with his youngest and most energetic wife. This is just a test, Ed said. There are passages, he remembered, somewhere in St. Augustine, about the necessity to cross over one's beliefs—by accomodating one's desires—in order to be effective for the task of spreading the word. These memories gave Edvard spiritual solace—but there remained the demands of the flesh to overcome.

The igloo, although damp, was surprisingly warm; and the offered wife (having done this many times) was skilled but undemanding. But she did smell—of fish oil, bear grease, and her own unwashed body. Washing was not a habit for her people: The winters were too cold, and in the summers when the ice began to melt, no-one cared. Also, washing could be seen as an affectation for the youngest wife of the senior chief, a sign that she perhaps was interested in the stories about habits of the people down below, stories that came by way of sailors from the semi-annual ice-breaker. Such perceptions she knew she needed to avoid—there were many younger applicants for her position.

Father Ed, quivering from the cold while at the same time examining his conscience, did sponge down a bit before he took his parka off. He ruminated on the logic and imagery of the matter, and remarked, surprised, on how warm his gonads were, even when washed in icy water. He decided then that he would be willing to engage in the offering of, undoubtedly primitive and ungainly sex, because it would create an inroad permitting his deeper faith to be spread out among the natives, his new congregation—a more potent gesture than when he first arrived and spread his offerings of rosaries and prayer books on the rugs of caribou skin—before the time he was invited to enter the communal lodge.

The smell of the youngest wife—her name is Umma—became the most important focus of his mission. To only first endure Umma's emanations, then in time to like them, and finally to wallow in them (the place where he lost temporary control of his soul) would be, he was sure, a portent of success for his mission. He began to think of the smells poetically—as a landscape for the nose, a promenade of whiff and sniff, each variant replacing the previous one with yet another—more pungent and unexpected than those so far reached—but unveiling the vast complexity of fecund nature. Father Ed (he told himself) you need have no shame—you have done all that can be done to reconcile the separate needs of soul and body, and you have done so in a way reminiscent of the travails of the ancient martyrs. Edvard knew that there had been others—missionaries, visiting chiefs, itinerant whale-hunters, and so forth—who in the past who had availed themselves of Umma. But none of them had the gift of belief aand the interest in approaching divine savor that he discovered in himself. He was uniquely able to go—as Umma (in her diagrammatic English) told him—beyond the Northern limits of Western taste.

Edvard did get the clap—probably as a result of Anglican infiltrators into Jesuit turf. He told the chief about it, who sent him to a medicine man, who gave him a piece of medicinal sealskin to wrap around his pecker, and told him to wait it out, to keep his fingers away from his eyes, and lay off the chief"s young wife for a while, because she is the most likely source. But Eddie was not happy with native medicine—so when a steamer came by, he went aboard and paid the ship's doctor a tidy sum for a large dose of antibiotics, enough to cure both him and her, so that they could pursue the sniffing that overlay the grunting and groaning—that had made their fucking so delightfully incomplete and lengthy.

But despite discovering his new interest in the smells of, e.g., ripe pudenda, Edvard knew that his missonary duties was just beginning—he knew that after the formalities and first niceties, there was serious work to be done: The village was crumbling, the hunting and fishing had become increasingly scarce, and the elders were putting pressure on the chief to do something about it. In his student days, Edvard had read Savonarola, also the writings of Macchiavelli and Kissinger. He knew about scapegoats, deflections, and the quixotic direction of public rage. In his specific situation, however, he did not know whether the chief was angry about his prolonged dalliance with Umma, but he did suspect that it could become a public issue, a target of blame for the village ills, and a reason to cast both him and Umma out to sea on an ice-floe in the darkest night of winter. There was too little winter wood for the alternative—an old-time burning at the stake—and anyway, the elders knew the limits of magic, and did not want to overly anger Western largesse by a too-blatant primitivism. But they also knew that the condemnation of illicit sex is a pretext the wider world has used for centuries as a way to get its way—and, in this case, setting the miscreants adrift on an ice-floe would be a non-inflammatory way for them to go away—especially when it's too dark and cloudy for the helicopters.

None of this had as yet become an actual threat, but the Thomistically trained Edvard thought it all too probable—so he made plans. He would eventually have to relinquish his calling—that he knew. And Umma knew that she wanted no one else than Missionary Eddie—not only because of his unusual ways of making love, but because she felt that he could take her to that lower warmer world where she really belonged. They spoke a bit about washing—which was not a problem for her, as she hadn't washed from childhood, and didn't really notice bodily differences through smell—although she was very good at smelling the approach of storms and the differences between spouting whales—but she knew there were these other differences that counted in the lower world.

They, Edvard and Umma, now agreed that they must quickly leave the village, and that their escape— the escape from their village down to Shishmaref had to be well-planned. Further, that they had to find their way through the low-country places until they found one where they could rest and plan more broadly. Life, rather than smell, was now at stake. They could not take a boat, for the villagers would be watching. So they quietly bought a sled and some dogs fom Umma's uncle—for an exorbitant sum but with a vow of silence—and on an early Sunday morning, they high-tailed it through the passage that separates the coast from the inner regions. Umma knew about such things, and she gathered the skins, the blubber and the oil that would sustain them until they reached the larger inland villages.

Although their departure was clearly scandolous, they were not pursued by the cohorts of Umma's husband. In the village, upon news of their flight, they became officialy regarded as dead—but the actuality was of little concern to the village politicians. Umma and her pale-skinned, politically maladroit, spiritually naive and quite ugly lover had, of course, died of natural causes—buried by an avalanche, ripped apart by hungry bears, or frozen to death when the dogs could no longer move—it doesn't matter now, as long as they're gone. Also, the council elders had agreed that the two of them are to blame for the recent ills: The flight of seals, the impotence of the walrus, the indifference of the birds. Why not blame it all on them? It would take the pressure off the ruling council. But, as it stands, Umma and her preacher are gone, we cannot kill them, the weather is getting warmer; and the fishing and spawning, in their absence, will soon be better.

Edvard and Umma reached the far out-posts of Western Civilization in poor, but not desperate shape. He had not lost his credit cards and so could pay for a place to stay; more importantly, he could contact his order to send funds for his return. He did not tell them that he had lost his faith, or that two was the actual burden for the price of his return. His superiors complied with a check—barely adequate for one—but they did not, at that time, voice their disappointment in the failure of his mission. Yes, failure! They knew it—and they knew he knew it. They had inferred this from the obsequious tone and the omissions of hard fact in his letters to them. Jesuits are trained to be particularly good at reading between lines—at giving content to the unsaid—so as to more precisely spot the intrusions of the diabolical. But father Ed was not giving them much of theological interest. Despite their long tradition of interpreting confessions, Edvard's superiors did not penetrate the real nature of the dissimulations coming from their emmisary to the north. He laughed and cried at what they could not know. Of course, they read the pap and blather of his reports—but how could they understand the deep-down sources of the trouble? How, indeed, could they dredge up into rationality the seductiveness of funk, the allure of over-ripe armpits, the glory of an unwashed ass-hole—and reconcile them with the needs of a holy calling? And could they imagine that all this sin was festering on a sealskin, in an igloo beyond the Arctic Circle?

But the fathers must have sensed Edvard's needs in abstracto—after all, he was one of them—and so they ignored, for then, the possibilities of the unmentionable. They did send him some money—although less than he had aked for. But, despite their secret hopes that he would find heaven before returning to America, it was not the last they heard of Father Ed. When he eventually did show up—after some years—he was wan and incoherent and fit only for a warm place in the Abbey corner. He did not tell them about Umma—it was not clear to him that she was real. He did not tell them about his leap from faith to disbelief—it was clear to him that this would not be wise. But he did make noise when he was hungry, and he often had to be held down when his urge to shout and stamp would reappear—which happened every other day or so. It is not possession, the fathers said, it is merely schizophrenia. None-the-less, while aware that they were stretching compassion beyond its acceptable definition, the Jesuits followed an older impulse of charity and forgiveness, and they did prolong his life.

Edvard discovered in the igloo that he had lost his faith—but only, as he insisted, the faith that was found in the dogmas of a celibate, codified religion. His spirit and his mind, his heart and gonads, began to go in different directions—and, to his initial surprise, they did not want reconciliation. He could not, any longer, attribute his buzzing in the head to his guardian angel—that timid creature had left him long ago. But the holy spirit—Ah, yes! Spirit, or something like that distaff God-part, was still there—changing as he changed. Why question its existence—look at the evidence: Something was making him jerky-dance instead of walking, and sing in abstract sounds instead of talking, and sniff for new positions instead of fucking in the same old way. It wasn't just him—an important part of the Trinity cared!

Umma took all this with her usual good nature—she had seen it happen before—in her village. Cold and wilderness invite possession by malevolant spirits: The incessant winds become voices demanding a response; ice-floes turning into incubi promising ecstasy and warmth—if only you discard your parka and pull your pants down to where the hot-cold-lips can reach you. Every year her village would lose a few to the rapture—as they called it. She had seen the possessed stumble into the far mists while laughing and wailing—never to return. The others who were not infected did not try to bring them back.

By now, Ed and Umma had gotten further south. They used the church's money to buy food; Eddie shot a deer (his first) and fed most of it to the dogs—but they did manage to barbeque the saddle (Umma knew how) over a small fire—and it made delicious eating. They also found some boots and ointment for their frost-blistered feet at a trading post. Eventually they reached a small town; it was larger than a village—with a mix of natives and whites—miners, hunters, early speculators, store keepers, local officials, a post office, and a few churches—it looked like a good place to stay awhile. But in that town, like the seasonal mud beneath the street-boards, there was also boredom, apathy, anger, and lots of alcohol and violence—the dark ingredients of loneliness.

Edvard had become quite interested in this mix of self-sufficiency and loneliness . He remembered an old Canadian movie called "Black Robe," a good gritty film about missionaries (Jesuits, actually—hence the "black") in the far north, with good and bad indians, killings, love with native women, loss of faith—all experiences he was now having. The film' s characters included an old Jesuit who had been scalped, his face hanging in layers from his head, and who was living the last part of his life tending to Indians dying of smallpox. Another Indian, who had helped the party escape from the bad ones, brought them to a clearing which he then recognized as the place where he was to die—although he was not old. He bade the rest farewell, wished his daughter happiness (although he could not understand why she had fallen in love with a paleface who was so ugly). He then walked away from them, and turned to stare intently at some rocks and overhanging trees as though they were childhood friends whom he had found again and could now die with to complete the circle of his life.

Edvard knew he had to do that—but in a different and more public way. Father Ed had been a preacher and a teacher. Now he was a dancer and a talker of tongues, a haranguer without a god to back him up—but he was amusing and would not stay still for long. It might be that he could form a group, a new congregation in that small town with folks who would dance with him, talk to each other in their remembered tongues, who would conflate the orgiastic with the scholastic, and get closer to the center of their living—however old or young or drunk they might be. One can invoke the gods, he said, without having to choose between them. This, then, was Edvard's project—his real mission, his rationale and task for what would remain of his own life. Umma was there—he knew—to keep him from falling into the pits below his foot-steps, and to divert, with her own charms, the anger that would be levelled at any controlling outsider who wanted to make the town his own creation, his work of art. Umma had now become devoted to him: She had learned to wash, although he now would neglect to even scrape the dead skin off hs feet. His love-making had once been an inspired epiphany of contrasts—but it had recently become a quick-stop on the way to the saloon. Yet, she continued to see him as a symbol of her freedom—the very thing her parents, and her school, had warned her against: If you reach out there beyond our truths, they said, you will end-up with the crazies on the drifting floes. You will then have nothing but the cold—the crazies will not help, indeed, they cannot help themselves, and we will consider you as dead. It will be an abandonment of your life if you choose this pale-one over us. Remember—the igloos here are warm in summer and not too cold in winter—better to stay with us.

But Umma was not like her kin—she had learned to turn her mind in directions diagonally across their straight line of past and future. In Father Ed and his ministrations, she saw a way out of her family's concoction of ancient fear and suspicious conformity. It did not matter what Eddie preached or how he fucked, he was her way to another of the selves she found she had—one of the many compartments he had opened up for her. So she went with him and did for him whatever it was he wanted but couldn't do. Sometimes, in passing, he did what she wanted him to do for her.

Father Edvard saw the action in the saloons as his first portal for entry into the world he was envisioning—for the villagers and for himself. So he went roaring there on Friday nights, and after a few introductory drinks, taught them how to dance and howl; and he assured them that the sounds they made were better than prayers, and their contortions while going for another drink would keep them moving into next-week's Sabbath, and also ease the way to paradise—when it came their time. Eddie (that name was easiest for them) did not call his pronouncements a religion; rather, he called them reasons that the Spirit gave the world—namely they themselves—to make a festival—a celebration of being here together, all alive, drinking hard, dreaming dreams of great veins of gold, replaceable loves—and bearing, happily, the daily burdens that would enable them to congregate next weekend here again.

Some towns-folk objected—pasty Anglicans, they were. They were egged on by the local clergy, who saw this charismatic interloper as a threat to their religion and their livelihood—perhaps, as they whispered, he was the the very devil in sealskin. And they were right (about the threat and sealskin). But the bully-boys in town were tougher than the locals, and usually more drunk. The Devil was no great shakes, they said, beside a hungry polar bear. These boys would come to town in waves—fresh from hunting or scraping the tundra for a little gold—just enough, you know, to keep them going. They found in Father Ed—not excuses but reasons—a fuller shelf of them than the locals could give—for the cruddy lives that they were living. The new rituals that father-Ed had shown them—the dancing, singing, farting in arm-wrestling contests, running out to pee novel patterns in the snow—these quickly replaced the mundane directness of their ordinary world. Everything that they did was now right—they could see that what they did was about the way they are.

Most of their choices of making things from other things (especially in the beginning) centered around a vein of gold and barrels of whisky. But slowly, more distant associations began to come together with the ones that were simply on the bar or in the ground. They all had their memories, but did not like them very much. Father Ed taught them that their memories need not be there only to blame themselves with. Rather, memories could be like spices—to turn the Mulligan into a Bouilliabesse. Eddie talked to them a lot about the varieties of French Cuisine; they responded, bashfully at first, by describing their most memorable moments of feasting after making love. It soon became clear—for Father Ed and his new disciples, that memories are a proper part of daily living—and everything (including anything and nothing) was now about something other than what they once thought was the way they had to go. Father Ed gave the Bully-Boys the gift of circularity. "Had-been" was now conjoined with "will-be" as is a spider spinning it's web around the past and present. They need not go anywhere else but here to make both past and future possible: "Bartender—three fingers of redeye—the long way!"

Edvard, after the spate of haranguing and dancing that introduced the first portent of infinite living to the locals—the small insistent sounds he first introduced that developed into the competing cacaphonies of other ways to go—pulled back a bit. There is after all, he argued with himself, a balance to be maintained between novelty and the ideas already there—in this sweet village. He then cautioned his friends, his boys, his parishioners, to choose carefully between the various options that were realistically available to them—and not to be seduced by all the ones now simply flittering around their heads—none of which, as he hastened to say, were really given them by him.

But that last admonition was a downer—"real" was a poison word—it just made the congregation mad: You let us loose, they said—OK—thanks for that, Eddie boy—but now stay the fuck out of the way.

The scarcity of women was a major problem in the town; there weren't many available ones around, and those few projected no semblance at all of sexuality—the town-women, young and old, were as constricted as sacks of rice and colder than the enduring snow. And everyone also knew that the few whores in town would give a man the clap. Umma avoided the sexual agressions of Father Ed's now waning congregation. In the main, the bullyboys continued to regard her as a spiritual priestess—immune to the world's desires and also too dangerous to be approached.

But there was one night when this understanding did not carry far enough, and the situation became lethal. Father Eddie was sacked out in an early drunk—but it was the night, as the wireless said, for celebrating Eskimo Liberation Day, and that was one reason for Umma to go to the big saloon. A deeper reason was her conviction—made, as was her way, in a period of stillness—that Father Ed was going nowhere deeper than the bottom of the bottle, and all his rantings about a universal ecstacy that would replace organized religion, was mere confusion between what he wanted and what was happening. In another stillness, some days after the first, she decided that it was now time for her to leave him—not just yet, but soon. At the saloon, Umma was greeted with cheers and ribald conjectures about how good she would look without her sealskins. She replied with her usual unreadable smile, gave a short speech in polyglot about the plight of the Eskimo people. After finishing her presentation—which few drinkers could follow– she gave them all a provocative wave, and the slightest shimmy of her hips.

There was a man in the saloon that night—a large hairy powerful primitive man—who had been in the Northern wilds for much too long, and had come back to see what talking has to offer. But what he really wanted was sex—enough of masturbating in the wild, where he couldn't get his fantasies of women to separate from the realities of animals. Umma looked to him like a large snow owl—bright piercing eyes, a hooked nose, sudden quick movements. He did not know what she was like under her skins—but no matter, he wouldn't need to see the rest of her when he was on her. He moved slowly across the floor and stood in front of her, shirt open, weaving and making ribald noises. He smiled, thinking that his size would do it. But when he crossed the line and reached for her, he felt a sharp pricking pain in the skin of his belly. He first thought was to scratch but then he looked down. Umma had implanted her fishing knife a slight way—a quarter inch or so—into his belly. She looked at him with her bright eyes. This is how, she said, after we catch the whale, we strip off the blubber. It takes two cuts, and much pulling.

Pay attention now to what I do! She slowly traced the cut from the front to the side of his protruding gut, moving the knife through the matted hair until the cut reached the wart that separated front from back. He looked down and watched the blood trickle slowly from the cut and settle into his exposed underwear. Umma then said to him: You are big and strong and could kill me with a blow. But I am faster and you are drunk—and the knife will be in your innards before you could raise your heavy arm. The big man looked beseechingly towards the bar, where all attention was focussed on the scene, and waved his hand at them as if nothing much was happening.

Drinkers, even in life-theatening circumstances, are passive—they know they cannot match the speed of the sober world. The big man knew this when he turned again towards his bloody belly. But Umma and her knife were no longer there—she had watched his eyes, as hunters do, for the moment they lose focus and turn toward the irrelevant. As soon as she saw the whites, she left. No-one saw where she had gone—and if any did, they were not saying. The big man, after a few more pints, went back into the bush—where sex is more of one's own making.

Umma returned to the cabin where she and Ed were staying. He was sobering up and practicing his gutturals for the next service. Good that he was there, so she could tell him directly that she was leaving, going back to her village where, given what she had learned from him and others, she could help raise her people out of poverty and the monotony of their lives. This is because she knew there is a moment when something can be done—not before nor after. She thought, for her reasons (which we do not know) that this is her moment. She wanted nothing further from Father Ed—except that he not follow her.

Edvard, strangely—after all their journeys through smells and spirituals—felt little about the separation. He did not try to talk Umma out of leaving, or stop her by insisting that the dogs were his, or warn that she would be killed upon her return. He just sat on a bench and watched her go, then walked back inside before she was out of sight. He never heard from her again, nor did she contact anyone to relay to him her later story.

Father Ed had lost his capacity for pertinent attention. He could no longer preach nor practice the sublime in his effort to halt the terrible times by carving a shape out of passing ideals. Umma had led him close to it—had given him reason to do it right. But without her, the bits and pieces went their separate ways, and he followed along—any which way was better than the prospect of staying.

When he got back to his monastery, he put back on the clerical collar that had been infused with the socks in the bottom of his dufflebag. No-one at the monastery thought much about his loss of faith—it was not unusual—especially for missionaries. Putting on the sweat-stained choker, after all the time he had been denying its significance, was—for Edvard—simply a gesture to get him money, a place to stay, and perhaps another mission. But he also realized that it was a rejection of all that had once been important to him. He knew that losing his faith was much like Umma's departure—it ended his world. Afterwards, he did not any longer care about his direction. Perhaps that's why she left him. She knew that he would not make that gesture; she realized Edvards limitations in the matter of belief as well as his avoidance of the scary state of non-belief. She, herself, preferred non-belief—but this could not be discussed with crazy Eddie -- so she left.

The story does not end here, but it peters out through a pulling back of my own attention. Attention, as I have said, is the proper response to matters of great moment—to transcendental constructs that surpass, through their surprising clarity, the world's wafflings in the muck and mire of ordinary life. Or, if such constructs are no-where to be found in our world, attention can create them anew, by compressing the lazy particulars into another, more noteworthy, shape—whether this be by song or dance, clandestine skills of imitation, or worldly plays on novel utterances. Attention is a force—the creative force. It is our commitment to what would otherwise simply be an unremarkable passing of the time of our lives.

I regret that I did not follow Umma, whom I love, back to her village.

XVIII
Walter

WALTER WAS ASTONISHED WHEN he learned that he is in the bloodline of the notorious Dracul—the ninth-century Transylvanian tyrant called "Vlad the Impaler" by his subjects because of his passion for sitting dissidents upon a sharpened stake and then adding them to the ring of impalees that marked the borders of his kingdom. Despite this heritage, Walter had no such tendencies. He was a gentle man, a conscientious academic and a self-effacing poet, who found his greatest thrills in teaching Blake and Keats—sometimes G.M. Hopkins—to undergraduates. His students were mostly engineering and business majors who smiled politely through his course because it more easily satisfied a distribution requirement than, say, 19th century philosophy. Despite the ease of his position, Walter was not happy with the way things had turned out in higher education. But were he to have thought of exacting vengeance for the creeping mediocrity he saw everywhere about him, he would not have followed his fierce ancestor and directed his wrath against those immediately at hand, his students. No, he would have mounted an attack upon the deans and provosts, those uncreative types who always strive to make the university responsive to changing times.

Walter was fortunate, however, that he lived in modern times, for had he lived in ancient Transylvania he would have been among the mendicants who tried to educate the peasants until old Vlad stuck sharp poles up their asses. That Vlad, of course, had many troops as well as a fondness for others' pains. Walter had known little pain and did not fathom power, only ideals, and so had few friends, even among the English faculty, who might, on rainy days, help him skewer central administration—had he been so inclined, that is. As it is, he had become content to leave the campus after class on Tuesday and Thursday afternoons, and walk with his notably erect posture and leather briefcase across the athletic fields, past the drugstore, and to the home he shared with his wife who also wrote poetry. Once there, he would check his mail for important literary letters and then ascend one flight to his library and the waiting tea which would be poured after he unlaced his shoes.

Walter had found out about his linkage to the Dracul line when, in a moment of whimsy, he answered an ad sent by a church in Utah which augments its revenue by tracing the lineage of almost anyone who subscribes to the service—one need not be a believer. Walter had heard from a colleague, whose own family tree includes British royalty, that this service is reasonable and quite reliable, and so, what the hell, a little fun, let's see what they, ha ha, dig up.

The information came in bits and parts, and for each new bit he had to pay a little extra; but the linkage was systematic, with a branch for each preceeding generation, honest question-marks where records were obscure, and upper-case for particularly important ancestors. Walter learned that he was the great-great-grandson of a noted Balkan painter, one Vlad Draculski, who fled to the Netherlands in the eighteenth century because he had angered a powerful courtier by painting two portraits of his wife, each portrait revealing a different aspect of her personality.

But Vlad's children and grandchildren—as they had been mothered by cheerful, blond, Dutch women—took a dim view of their artistic heritage, and instead became merchants, bankers and the like, professions which eventually brought them to the new world, to New York City, where Walter was born, the son of a broker and a home-maker. His last name, van der Vlad, attested to this wholesome ancestry. But unlike his siblings who were blond and cheerful, Walter was often dark and moody, with little head for business and a seemingly innate disposition to becoming a tormented soul. This combination led naturally to his interest in poetry and also kindled questions about his origins. At the time when he began to search for himself in the family tree sent him by his western apostolic friends, he realized that he must bypass the intervening generations, however evident their claim to him, and find another way to his mediate ancestor Vlad, the one who was key to his authentic line. Walter felt a strong affinity with that rascal, as if he himself were the son conceived by the serving maid whose timely warning had saved Vlad from the gibbet but doomed him to the artworld of bourgeous western Europe. Or perhaps it was the countess who bore him, which would explain the indifference of the count to him throughout his childhood—for it was whispered that his was a six-month gestation, if one figured from the time the count returned from war to find the dual portraits that the rascally Vlad had painted, the naked version revealing a mole that none besides the count should know about.

Oh, Walter thought, he would love to have been Vlad's son! This was no idle fancy, he assured himself, no counter-factual soporific of the type he primed with a martini on Friday afternoons, when the weight of his vacant students and his timid poetry grew too much to be borne directly. This was a beginning! By reaching across the documents to his great-great-grandfather, Walter had rescinded his willingness to continue plodding across the weekend to Monday. No; today he would use his memory and his fancy, seamlessly combined, to search for the gate that leads away from ancestry and into archetype. Once found, he would flee through it, into the unspeakable perversities which mark a different line of intervening years—begun by those who are his true source, and whose dark loins would henceforth receive all credit for Walter's better poetry. Perhaps then, having braved this much, he could overcome the fear of sharpened stakes, and venture even further back, into those stormy mountains which decent history has avoided, and plum the introspections of the Impaler himself—his source of sources—where, amidst the stench, the overeating, loud laughter and red rage, he might find a true portent of his past. That, he thought, is the way to good poetry!

So Walter forsook his present clan right through the father of his father's father, right through the last documentation of the loving and supportive ancestry that had given its best to what the new world offered, and had succeeded so well at assimilation that, except for progeny, they passed unnoticed through the century and a half that followed the arrival of Vlad's sons and daughters upon these shores. Walter forsook them all because he knew—as befit his retrospective journey—that he would not find himself until he left the linear path at the gate, and purified himself through storms and suffering—until he reached the melancholy divide that separated the painter Vlad from his clan. This is the Vlad who said of himself that his life ended when he saved it, the same Vlad who had left his best part in the innards of the countess (It had to be the countess, Walter thought—not the serving maid!) and thus produced an early line divergent from his later issue—the line, Walter was sure of it, to which he really belonged—although the ancient name 'Vladimir' had been twisted by later progeny into the American 'Walter.'

There you have it, a map to the revision of his past that would give Walter the chance to know himself, to ground the more grotesque of his affections—the ones he could not face even in the privacy of his daydreams—within the steamy soil of Transylvania. Sharing common ground with the young Vlad gave Walter the courage of a true perspective: From this point on, he will open his daydreams to the images of desires he had hitherto acknowledged only by their names—and more, he will expose them, later on, to the public through his poetry.

Walter's wife had been his student during the time of his most popular teaching days, when he encouraged all his students, even the dishonest and expeditious ones, to write poetry far beyond their earlier and later selves. Marthe—that was her name, Marthe Liebeswort, the daughter of a well-to-do professional couple from Westchester, who sent her to the midwest for her education and safety because she was their only child. Marthe was quite honest and sincere, although she was uncertain about where she stopped and started. She compensated for this vagueness by writing detailed descriptions of her outward self, and by wearing skirts which accentuated the soft curve of thigh following from her rather prominent and fleshy hips. Walter, sensing talent, pressed her for greater introspection, and in response she brought in fragments of her inner self for which she would request a grade according to the degree of introspective success.

To adjudicate her rate of progress was far beyond what class-room time allowed, so Walter and Marthe—the sophomore and her professor—extended the critiques into the campus coffee-shop, and when the regularity of these meetings began to attract attention, to more and more distant locations until sufficient privacy could finally be found in Walter's bed. Marthe was not discomfited by all this; she took it rather as the long-awaited directive for her life; and so she introduced discipline into their coupling, a consistency to their comings and goings, which lent stability to the nights and predictability to the days, and signalled the inevitablity of their marriage. Walter took his new steadiness with good grace; he rather welcomed the methodical fucking and the predictables of house décor and home-cooked-meals, and he could not think of anything specifically his own that was being interfered with.

But his teaching grew erratic—perhaps it was the years of repetition, or the result perhaps of his libido having become more circumscribed. He began to pick and choose among his students, abandoning some before the ends of their required poems, justifying his departure from their needs by his lack of time and their lack-of-talent. Paradoxically, as his students became more distant, he felt an increase in his own poetic powers, as though his maturing familial sense now made it easier for him to build on what it is he wrote about. Over the next years, he published in literary magazines, had his work included in anthologies, became a respected voice on curriculum committees, was granted tenure and appointed creative emissary to gatherings of the wealthy university community. He did not think to blame his wife for the direction his achievement took—for what was so bad? Would he prefer to dream upon a rock in Maine or write crap-copy in Los Angeles? Yet he would go into rages, almost every afternoon, after just a few sips of his second martini, and could remember nothing of the cause when sleep arrived; yet with each morning came a growing disaffection for the day that followed. Marthe, who remembered everything, never spoke about last nights or following mornings, and she did not help Walter see the pattern which had gridded over the vision of his life. More than anything she wanted the strands to remain secure. Carnal Love was made on weekends now, unless Walter's critical eye happened to seduce an errant graduate student, which event would lengthen the hiatus between familial fuckings from weekly to bi-monthly—so Marthe always knew.

Marthe no longer showed Walter her poetry although she told their friends that she wrote everyday—but she said the poems were too personal at this point, too introspective, to be read by others. Walter did not show her his, of course, but he assumed that she, like all his other friends, would read his poetry whenever it appeared in print. Actually, after that first class, Marthe had stopped reading his poems, for although she tried she never really liked them; they were mostly about words and she was interested in things; she had always been a bit dyslexic and could read just so many words on words before they interchanged. Earlier, Marthe had regarded this as an insufficiency, a secret limit on her talent; but later she came to feel that her poetry was better than Walter's because, in fact, she used pictures as well as words, and both of these were about things other than themselves. She found her pictures in her family album, in newspapers, and in the soft-porn magazines that Walter liked; she wrote words upon her pictures and she pasted pictures over what she wrote; the pages of her poems sometimes glued together in the humidity of mid-western summers but she didn't mind because, anyhow, she didn't plan to show them to anybody else. From the outside, through the glass of collegiality, Walter-and-Marthe's middle years looked like a version of the dance that all had tried to learn—a glimpse of practiced private steps and an occasional show of complex public turns. Discounting the possibility of suicide or homicide, they would have danced away that way until retirement day—had it not been for Walter's frivolous gesture of enlisting the Mormons to trace his ancestors to their source.

Walter and Marthe separated soon after this. It was as if the divine enabler had given them the means to break their iron pattern of excuses and just leave, with a smile and wave and time-of-day, no more, no more needed once it had been done. By constricting her bowels, Marthe could have handled Walter for the remaining years, but she had no stomach for those Transylvanian-Balkan spooks he increasingly interjected into their every conversation. And Walter found he had no cause for anger anymore, no matter how many martinis he guzzled in an afternoon. He would actually stagger out without the dog for early evening walks and he would look for faces in the darkened sky. He no longer pursued each day's small slights because he was busy tracing his less documented ancestors into the nights in which, through unspeakable deeds, they overcame the smallness of their ancient days.

Marthe kept the house and dog. She found she could no longer stand their academic friends. What was it exactly? Did the look they now gave her contain a smug suggestion of their greater fortitude, their success at constipating the calls of nature and declining the odds of chance? Marthe looked back at them: Were the dank canals that connected their various parts, sluicing over nose and mouth before spilling into the wattly foothills of the neck, an act of God or university-made? Were these crookly grooves nothing but the lower remnants of a project of high expectations—a dam to stopper an untamed river, now abandoned by its colonial power? Why had all these former friends become so ugly? Marthe drifted towards communities of singles, as found on all college campuses, who couple ideology with their situations and invest short spans of time into their unrealized hopes with great zeal. Marthe broke her pledge of secrecy and showed her poems to an intense new friend named Ann, who sat cross-legged and barefooted in Marthe's attic and pored over every scrap of paste and scribble she had ever made. Ann was a facilitator, a life enabler, who could have run the university had she not been small dark thin, and uncomfortable with fools. But she did match Marthe's poems with a publisher, and elicited a first uncontrolled scream from Marthe's lips by nibbling on her clitoris. After a time they moved to New York, where their talents found counterparts and look-a-likes at every turn and they had to work harder than they had known to just be noticed; but they succeeded reasonably well and were quite happy even though the others who had been there longer, no longer believed in happiness.

Walter had an old station-wagon from his camping days. Although it now had little left but a reluctant motor, rust, and balding wheels, it was an adequate carrier for city driving, and keeping it sustained his fantasy of piling all essentials into its large rear and going to where, suddenly one day, one had to go. So when Walter left his house and home, he had a ready-made short list of necessities and beloved objects, and it took him only a few hours to be out of there and on his way. He didn't go far at first—to an apartment in an Irish section of town, one of the ethnic enclaves in the city whose derogatory nickname was used by its own inhabitants as a way of designating a style of wedding nights and barroom brawls. Walter did not use the name; he was part of the classless enlightenment for which such names are unredeemable ethnic slurs—but he nodded and smiled when his neighbors used it in their morning greetings.

Walter could have lived elsewhere, closer to the students and the younger faculty, but he chose not to. He wanted privacy of identity; an equivalence of status between Professor Walter and Tim O'Leary the plumber who lived next door. This leveling was not a symptom of the ecumenical trots that often urges liberals with secure pensions to seek the realities of ordinary life. Walter's choice of location was a practical one, for he truly did not know what he would become when reconnected to the sighs by moonlight, no quarter in the battles, and the merriment in the castle that were his birthright. For the rawness of his actual lineage, he had to be prepared. He had to face his sources fresh, without Marthe or creative writing, indeed, without Professor Walter—or else he would be doomed—like the villager who having heard the cannon roar, ventured into the liberated lands without discarding the old ways—and was blown to bits.

There were now few women in Walter's life. The graduate students found the poet's lot he presented, too lacking in advanced subterfuge and intimations of the larger world—they would not now trudge to his one-bedroom on the third floor for cheap white wine and gray sex. And the wives of successful men who had followed him to the classes he gave at conferences in Boston and Montreal, were turned away—when they came to town—by the smells of cooking in the hall and the unwashed towels in his bathroom.

But there were more important reasons. Walter had discarded all that sex could be a compensation for; there was nothing in his life that needed filling—so there was no need to intersperse refashioned breasts and variously colored pubic hairs to frost a career and marriage as there had been before. Walter now had a mission; he was on a pilgrimage of the older sort—scratching at true belief—and his lusts could wait while he retraced the steps that his ancestors took, centuries ago, to save their asses. Now, he too must save himself, conserve his vital juices so that at day's or time's end, whichever comes first, he would be full of himself when he meets the countess—his one-time bride—for the first time.

Walter walked the path that grew steeper towards evenings and left him above tree-line when darkness fell; looking up he could see the creatures made by stars, coming to be tested by the cold and unremitting wind. He did not sleep but nodded through the night before his small fire, and the fierce eyes that circled so as to penetrate his edge of light, would retreat when his head jerked up as it hit his knees. He built his fires on the highest ground so that the circling ones, in case their hunger overcame their primal fears, would have to run uphill to get at him. But he also knew that fire is most visible on high ground and so each night he marked the sequence of his journey for the curious ones of a later time. The first bird-note that sounds just before dawn caused the circling eyes to dim and move away; and the warmer winds of morning came to show Walter the downward path which would lead him to the river.

The last night of a previous life was also spent on a river's edge. The eyes that had been following him were particularly fierce that night, punctating their flashing with growls at the prospect of gaining their prey. But they feared water more than fire, so Walter built his fire high and lay his groundcloth between it and the river's edge, and slept well past the first song of morning. The water's flow is gentle in the early hours although it picks up speed as the day goes on and becomes a torrent at dusk. No-one would dare to ford it during the night. Walter dreamt of a lay of stones which, if one stepped carefully, would permit a crossing in which the wetness would stay below the groin. But Walter was not careful—he did not yet understand the wilderness. He dawdled at his breakfast, and the water moved faster with each hour. But he was lucky, although his eastern roots had not equipped him well for treacherous stones. He slipped a few times in the crossing, and the water reached to just beneath his chin—but his head remained dry as did all the precious things he had piled on top of it, balancing, with his stiff neck, the best of his old life as he crossed. He was not sure that his new life would need the testimonials and reviews stacked neatly on his head, but he took them just in case, in case those he met would want to know his worth before he could prove it.

No sooner did he reach the other bank, when a horse whose rider brandished a glistening sword, came charging down on him. The sword swung and cut off his stack of testimonials as neatly as would a freshly sharpened guillotine. Perhaps the rider mistook the stack for his actual head and thought he had decapitated him; but Walter could not know what was intended, for with that mighty slice both horse and rider vanished in a drum of hoofbeats punctuated by loud laughter. Walter let the scattered papers blow; he could not use them to impress these people, that was for sure. But the laughter (was it the rider who laughed?) troubled him more than the danger to his resume or his head; he had heard that laugh before. He remembered that every time he reached a new beginning in his life, the laughter would sound—when his Buster Brown hairdo was cut short at the age of seven; when his mother sewed into the night to peg his pants when he was twelve; when, at sixteen, big Sylvia wanted him to stop looking at her breasts and levitate with her; when he tried to find out why he was not invited to the annual writer's ball; when he was spotted by students as he picked his nose in his convertible at a traffic light—and they laughed until it changed.

At other times, Walter did not mind the shame too much; he knew the ones who laughed—dumb nullities they were, small-minded segments of intestinal worms, killers of butterflies, ugly fucks. But for this rider and his sword, he had brought little with him that he could use; he crossed the river to take on what his ancestor Vlad had left behind—so who laughed this time?

The path leading away from the river was steep and narrow and could be travelled in only one direction, but it was soon joined by another path and then another, and with each such joining it grew wider until it became a road of sufficient breadth to give Walter confidence that a center—a town, perhaps a castle—would soon be reached. But Walter was not happy as he walked. The ebullience at having so easily forded the river and having so deftly avoided the assassin's sword, had vanished. He could not forget the unwarranted laughter. Also, there were no other travellers to be seen although the way was well worn; and at each juncture where paths came together and the road would widen there was an upright pole, taller than a man, with a pointed apex stained in various shades of reddish-brown. At the base of each pole was a plaque on which was inscribed what Walter took to be the name of a sin or crime. There were a few poles labelled "blasphemy," and "misogyny," but by far the greater number were named "tax-evasion," and "money-laundering". Walter assumed that the iconography—for these poles were evidently art-works—harkened back to the time of the great impaler, the infamous Dracul who was also father of the country (and perhaps the founder of Walter's line). But at the same time he was surprised at the naive realism—the fake blood-stains and phony labels—particularly as these works are on public exhibition and should therefore show greater accuracy in their repesentations.

The road Walter took led to a castle not a town; but Walter's hopes that the castle would exemplify the spirit of enlightenment—the new version of himself that he had come to find—these hopes were disappointed. The castle was a thick grey dank affair, with misshapen turrets, an overgrown moat and one dim light—right out of movies he disliked. Only an intermittent scream, he thought, was needed to make it a major attraction for a packaged tour. Then he did hear an actual scream—once, twice, three times—followed by the blowing of a whistle. The footbridge across the moat was down, and although he was uneasy, he walked across to the gate, a rusty metal one, upon which hung a sign saying "VLAD'S CASTLE—hours 1-4 on weekdays, noon-5 on saturdays, closed on sunday." The day was Sunday but the gate was not quite closed, and as he pushed against it he heard, in addition to the squeak of hinges, the sounds of conversation—but fortunately no-one appeared to challenge him. The screams had come from the building in which the light was shining, and Walter thought it reasonable to go in that direction for, after all, he might be quite mistaken about the content of that noise. The door to this building, made of a rough-hewn heavy wood, was also not quite closed. Beyond it stood a turnstyle, but Walter's early New York days had taught him how to jump over turnstyles and he did so now, thinking it much too early in the adventure to commit himself by passing through.

A long passageway of damp and ancient stones led to a large chamber that was illuminated by torches jutting from the walls. Walter's footsteps were quite loud and he was surprised at this, not knowing whether it was his clumsiness or an echo that made the noise. But he was not heard, for the chamber was full of bustle and loud voices, and so Walter passed unnoticed through the last archway to see a most remarkable scene.

A rack—yes, the medieval instrument of torture that sends unlikely souls to their destiny—was positioned against the far wall. Upon it, for all to see, was stretched a young woman with light red hair just discernably darker than her pink and lightly freckled skin. Her gown was elaborately embroidered suggesting wealth and high station, but it had been torn open at the bodice, and her breasts and navel were exposed. Strands of pearls circled her neck, but her feet, held fast by chains, were bare. A voice from an alcove off to the side said something in a language Walter did not understand, whereupon a tall bearded man dressed all in black strode to the center of the room, pointed a long stick at the woman—who had lain quite still thus far—and loudly spoke in that same language. Then the wheels and pulleys of the rack began to rumble and the woman began to writhe and scream in close counterpoint with the bearded man's sonorities.

After a goodly time of screams, rumbles, and invectives, the voice from the alcove was heard again and a group of men with spears and armor charged into the room led by a youth in a flowing shirt and buckskin boots brandishing a rapier which he ceremoniously thrust through the body of the bearded accuser. The others released the woman from her fetters and lifted her,—somewhat abruptly, to her feet. Curiously, she showed no signs of the broken bones, displaced joints, and torn muscles that serious racks are designed to cause. Once standing, she proceeded to straighten her hair, took some pretty poses, then pulled demurely at the remnants of her gown. Her savior, waving his bloody sword now ran to her, and accompanied by a general brandishing of spears and loud huzzahs, they kissed and hugged and danced around the rack. When the festive noises began to wind down, a short round man in a beret came forward clapping his hands and speaking jovially in that outlandish tongue, whereupon the bearded inquisitor rose nimbly from the floor, the young couple stopped their groping, and the opposing forces stacked their weapons and wandered off.

Mesmerized by the implausibility of what he had just seen, Walter stumbled aimlessly around the chamber but soon attracted the attention of the man in the beret. To a query directed at him in the local language, Walter responded in English that he is a professor and a distant relative of Vlad's and he is here to seek the countess. "Ah, American," the round man said. "There's a special group from Miami scheduled to come through this afternoon, that's why we're rehearsing. My name is Emanuel—they all call me Manny for short. I'm the director over here—although I did my early work in Los Angeles. You're looking for the countess? She's not on tonight. The red-head you just saw is our other lead—isn't she something. The schedule ordinarily puts her on the rack monday through wednesday; what with her good manners, long legs, and flawless backside, we all thought she has the perfect temperament for her part.

But the countess draws a larger audience (she's the real thing, you know, and she really owns the part she plays). She performs on thursday, friday, and saturday; it's a flagellation scene, more esoteric than this, more nudity and some soft-core action. Very zesty—but we're having some problems. Her regular tormentor quit on us today and we're casting for another, but we've not had much luck so far. Torturers are hard to come by—the character is complex, both sadistic and solicitous, powerful but with a vulnerable edge. Not many like that in this back-water—the locals don't really understand the difference between fact and fiction, and they wander across the edge in both directions; this causes dicy behavior and bad acting. But I have to keep looking. Actually, you know, I've been looking at you as we talk, and thinking (all preliminary, of course) that you sort-of look the part: You have good posture with a high waist (good for looming), soft but devious brown eyes (never wholly truthful) and a tremulous mouth with a curling bottom lip (good for sneering and cajoling). Your hands do shake a bit, and you shuffle a lot. But that's ok—good torturers are like that. Want to try for it? We'll be reading in the morning—Walter's the name, you say?—nine o'clock."

Walter walked quickly back towards the town. What a stroke of luck to find the real countess in a theatre of make-believe that, like all the local art, is seemingly in the realist tradition! He had no doubt that he could do the part; the years of lecturing about historical abominations would serve him well, and all he needed was supper and a good night's sleep to prepare him for the morning. But as he continued walking towards the inn, he heard hoofbeats on the road behind him; turning, he saw a rider bearing down and swinging his sword, the same one who earlier had separated him from his resume. The sword swung, a head rolled upon the ground, and the rider disappeared into the dust laughing the laugh that Walter had heard before. Oh my God! The rolling head was his! But that could not be because he was looking it in the eye—and it was looking back! The head had come to rest in a puddle of pink fluid, and as Walter poked it with a stick he saw that it was not flesh and blood but chicken wire coated with painted papier-mache—a crude although convincing expressionist effigy. Walter was of course relieved—he still had his head—which is now the subject of a representation! He was stirred in heart and mind. This whole place, he thought, lives on illusion. It is based upon a premise of deceptive reciprocation between artworks and real things, the same premise that underlay his own desire to interchange his present with his past. He had reached his destination. Now he felt at home.

When Walter reached the inn, he was greeted courteously by the rider of the horse. The man had replaced his sword with an apron, and was busily serving wine food and pleasantries to the somewhat impatient tourists from Miami. He confided to Walter that not every tourist is greeted by faux-decapitation—only the important ones. (Dircctor Mannie must have sent word ahead). The red-headed girl that he had seen stretched so notably upon the rack showed Walter to his room. In response to his compliments on her performance, she proudly told him that the actor who rescues her in the play is her boyfriend, and they hope these roles will soon lead to an offer from radio or TV—perhaps in America—and let them become artists fulltime. She then kissed Walter on the cheek and told him he has nice hands.

The proprietor, a garrulous soul, came to sit with Walter after dinner and reminisced about his grandfather who had been a captain in the last count's cavalry and served as executioner when the need arose. "The country is much different now, tourism having largely replaced war; but of course, it is the bloody history of the place—the legacy of the original and terrible Dracul—that brings the tourists here; and so we give them pageants of the olden times—torture in the castle three times a week, free decapitations for the prominent ones, and impalations performed at every crossroad depending on the weather. (These days, you know, its all papier-mache, a little paint, and chicken wire). But we find tableaus to fit all tastes, and some visitors like our work so well that they buy property and stay on. I must say that I prefer Americans to the others; Americans have a finer taste in both food and fantasy; and, you know, as artists we need an attentive and generous audience."

But for all his amiability, the proprietor was not pleased when he heard that Walter might join the pageant in the castle; politically he was of a protectionist bent and felt that foreigners should supply money but not compete for jobs. Walter then told him of his blood kinship with the land, his own mission as an artist, and his obsession with a legendary countess, his earth-mother, wife and sister, whose decendant might be the very one he will flagellate tomorrow. The proprietor sighed, the sound emerging from both throat and bowels; and his ruddy open face turned inward like a toad's. But this passed, and smiling he asked Walter if he had met the count. "The count is one of those among us who is not an artist—he prefers the real world."

When Walter left the inn the next morning, he again met the red-headed girl who was setting tables for the late breakfast. She took his hand but gave him a somewhat anxious look when Walter told her that he was auditioning for the torturer's part. She knew that her boyfriend would be jealous when he learned that Walter—a foreigner, poetaster, ugly to boot—might be first in line for a part he had long aspired to. The Flagellation Episode is performed in prime time—all the notables come to see it—great exposure. The main draw, of course, is the countess—that impossibly voluptuous woman for whose company, it is said, huge sums have been offered, and as it is said—declined. To play opposite her would go a way towards fame—and also give him a closer look and smell (even a feel if he was clever) at and of her exposed limbs than he could ever otherwise achieve. He would risk death, he muttered, to hear her scream into his ear. Evidently, this young man was a simple soul with little talent; that was even clear to him. But his lack of talent—and his being deprived, at this his most needy time, of realizing these insistent onanistic fantasies—stoked the beginnings of a murderous rage.

The scenery was being set up when Walter reached the theatre, and in the middle of things stood the director still in his beret talking earnestly to a women of substantial poise and girth. She was dressed in tan riding clothes which nicely matched her long unruly hair, and she tapped her boot—impatiently, as Walter thought—with a leather crop. When the director saw Walter, he immediately waved him over and then stood beaming, his eyes moving from the woman to Walter—with her boots they were the same height—and then back. "This is Maia, the countess Luciano, She is our star; and this is Walter—what's that second name? You don't know it? My, that's a new one!—but we're all first names around here, friends as well as colleagues; and I only pay in cash."

So, Walter, you'll be Maia's leading man if it works out; but from the way you two look together I'm sure it will—if nobody fucks up." Maia said nothing, and when she left she gave Walter a look which he could only interpret as quizzical. "While she's changing," the director said, "I want to tell you about the scene. Believability's the issue. Our audience doesn't want artifice or allegory, dance or declamation. They want the pain and blood to mingle in their groins, to be as real as their own sweaty armpits—and give them good reason to do-themselves for at least a month. They want to be Dracul's invited guests—fellow voyeurs—when he exacts sweet revenge upon the loves of his enemies.

"You'll be dressed in black, as are all our torturers; it's better in this case that you're clean-shaven, a beard might get entangled in your swing. Here's the whip; see, all latex, light as foam, with little openings through which red dye will stripe her back upon each stroke; swing it carefully so that it doesn't break. The soldiers will manhandle her—feel her up as they string her up, haha—but your pleasure is to be located only in her pain; you're supposed to be a kink—like Dracul, like the count, like our audiences bless them. Don't worry, Maia acts well enough to give you all the motivation you need. Oh, one more thing. You may see a short dark man with enormous shoulders and a scarred smallish chin standing by the door. That's the count; don't let him distract you, but mind that you act professional at all times. Well, good luck. Places everyone!"

The play began with Maia's entrance. She had exchanged her riding habit for a flimsy gown that was open on both sides to show the hills and valleys of her extravagant body. The soldiers dragging her could barely keep their footing what with Maia's grand gestures of dismay—but it was obvious that they enjoyed their work. They dutifully rubbed her breasts and pinched her buttocks as they fastened her to the post, then carefully stripped off her gown (it would be used the following day) while avoiding the expressive flailing of her legs. All the while she ranted, gutteral and sing-song, in the language of the play. There was a pause, permitting the audience to have a good preliminary look, then Walter made his entrance. He responded to Maia's rantings in a deliberate voice, reading from a phonetic list of phrases the director had given him. "Speak them in any order you like," he said, "they all are about sin and penitence, but have them memorized by Wednesday." Then Walter began to swing his whip, awkwardly at first, spattering a bystanding soldier, but soon he got the hang of it and felt increasing pride in the symmetry of the patterns he inscribed on Maia's muscular although fleshy back. She punctated his every stroke with a melodious response, and the air became charged with the tension of believability. Walter could not but notice the count perched like a still black boulder in the corner, but when the scene had ended he was gone—to relieve himself, as Maia later said. During the break, Maia complimented Walter on the quickness of his study but suggested that they, just the two of them, needed more rehearsal— particularly in the rhythms— and would he meet her at the crossroads in the foothills, by the double blue and purple impalation, tomorrow afternoon at two. Then she too was gone.

It was a long way to the crossroads in the foothills and Walter trudged uphill feeling unusually tired and a bit ill at ease. The pretty servant-girl had given him another of her uneasy looks that morning, and this time Walter took it as a warning. But of what? Was it a general forboding—as peasants often have—or did she know something, perhaps about the count and countess, or the reach of Dracul's legacy, that he—softened by the make-believe of university life—could not comprehend? When he reached the crossroads, sweaty and more than slightly ill, Maia was not there, so he occupied himself by examining the double impalation. It was made of wire and papier-mache in the same manner as the head that the horseman had rolled in the dust before him when he first came to town. This work was more sturdy, however, covered with a resin to guard it from the rain; and the realism was more painstaking, with the requisite agony and despair clearly etched upon the features, and the coloration modulated by a subtle sense of decay. Walter was distracted from his studies by a horse's hoofbeats; then Maia came riding up the path. Her seat was practiced; she guided with her knees while the reins lay loose, and she wore an all-white riding habit which enhanced her girth. Upon dismounting she tapped Walter playfully on the thigh with her riding crop and said he needed whipping lessons, but first they should fuck because she had been feeling horny since she first saw him.

No help but to do it, Walter thought, much flattered. Doing the dirty might mean his job, his chance of connecting with his origins—but, after all, is it not the countess who has filled his every tumescence since he first began his pilgimage? The act itself was not too difficult; Maia showed him what she wanted done, and Walter went along, grateful that the confusions of his recent past were mollified a bit—if only by the certainty of a straight-forward missionary position. After a polite post-coital exchange of anecdotes about their respective lives, and a date to meet again at week's end, Maia galloped off leaving Walter to trudge back to the inn, feeling even more ill, his eyes tearing as he squinted into the setting sun. At supper, the freckled servant girl looked at him compassionately, and whispered that she would see him in his room that night. She treated Walter's fevered penetration of her as if it were a necessary validation of the warning she had come to give him. "Beware of the young man, my boyfriend, who comes to stab you in the play on saturday; his sword is real and he will kill you as the count has told him to—a mistake, I'm sure, of course—but if he does not succeed, as he told me, the count will—on the road that leads back from the castle to the town." She then left with her soft sweet scent on Walter's face.

Walter's next meeting with the countess was much like the first, except that he did not feel as ill and she used her riding crop liberally on his thighs to demonstrate the finer points of whipping-rhythm, regretting as she stroked him that he could not reciprocate because the count would see the marks.

Walter lay awake that night hoping that the servant girl would come to him again with more dire warnings or just her sweet softness, but she did not. Strategies to avert his death staged a disorderly parade across his dreams, and the opposing cadences of fear and prurience so confused him that the twitches of his face and the spasms of his gonads lost all coordination. As dawn approached, he awoke sufficiently to think that he could probably avoid the soldier's sword—the young lad did not seem that ambitious and, indeed, his sweetheart might be softening his resolve at this very moment. But the count! Ah, that was another matter. The man is an exterminating devil, the ardent dispensor of unnecessary evil whose malign influence upon the world is limited only by the fact that he is rather ignorant, clumsy short and ugly, and had early gambled away the family fortune. So he is now reduced to watching his wife's great white body sway each Thursday through Saturday before assorted Englishmen and Germans, and then to get his jollies by preying on itinerant American poets in the hope that their tepid blood will quench his bitter and unfulfilled desires. How can Walter lily-pad, high waisted and too large of foot, escape such a one?

Saturday's theatre crowd was noisy and unruly, fortified to the full parameters of appreciation by the local wines so liberally served at lunch. The play began as scripted; Maia projected more passion than usual in her efforts to pierce the alcoholic haze, and the soldiers showed greater respect for the power of her gestures. Walter, however, was not fully in his part; the particular mix of fact and fiction now overflowing the script disturbed his concentration—his utterances were mumbled, and the stripes he lay on Maia's back somewhat chaotic. The audience, happily, was unaware of this— they were entranced with the play of her behind. Maia used these talents to maintain the play's intensity right up to the moment when the young savior, sleeves dangling and sword swishing, burst into the room and made for flagellator Walter. Then, everything broke down. Contrary to the script, and without regard for Maia's stellar presentation, Walter threw a stool at the young avenger. This tripped him up and permitted Walter to step on his ear while running past into the gloomy corridors in search of an escape from the greater danger of the count's pursuit.

Walter knew that death lay waiting on the road back to the inn, for at some dark turn the count would grasp him with those immense hands and subject him to the most unspeakable torments before he died. Why? Well, that is what the count does; that is his nature—as much as it is Walter's to speak of love in rhythmic imaginative ways. So Walter thought to outsmart the count by remaining in the castle during the night. He would tiptoe up the back-stairs that were lit only by the moonlight squeezing through the weapons-ports, and he would find an unused room in the higher reaches of the tower where he could hide among the cobwebs and other relics of the past. Then, at first dawn, he would seek out the road to the other side of town and follow it down until he reached the docks—and there he would book passage on a ship that would take him to a more prosaic land. Once safe from the deadly make-believe, the fraudulence of ecstasy and pain, and the reality of death, in the Impaler's domain, Walter could put aside his quest for ancestral wholeness and instead devise a way of redirecting his affairs to things that were less dangerous—even if more familiar, and so less satisfying.

High up in the tower Walter found a door with hinges that looked as if they would not squeak; he opened the door and slipped into a large room whose walls were lined with books. It was a serious room, with heavy drapes standing sentinel beside each window, and wooden desks whose tops were stained with ink and worn smooth by monkish elbows. At the far end was a fireplace in which a cheery blaze was crackling, and facing it were two upholstered chairs one of which was occupied. The count—yes, alas, it was he—rose and greeted Walter in a gruff but not unpleasant voice, and motioned him to sit in the vacant chair. Like a bird transfixed by the stare of a serpent, Walter did as he was told—numb even to the prospect that to obey surely was to die.

"We are relatives, you know," the count began. "Yes, you and I are descended from the same mother although from different fathers. You come from Vlad the painter, the one who fled to Holland to save his ass, and I am progeny of the old count Luciano, he who tried to get Vlad impaled—as was then the custom with philanderers. You get my point. That effete snob with a miniscule talent—he probably looked much like you—painted the countess in her threads and also in the buff. During one such sitting, your line was conceived just before the count returned from war and chased Vlad out of the bed and across the waters all the way to Holland. But once painter Vlad got there—and this is only because of the Dutch aversion to extradition—he was safe, and he then proceeded to woo other fat blondes with whom he sired the pot-smoking poseurs you can think of as your predecessors.

The count—my ancestor—disgusted that his blood-lust was unfulfilled—returned to that self-same bed from which Vlad—your ancestor—had fled; and then spent a fortnight assuring the countess that he was back, the war was over, and he was ready for the peaceful life. After seven months—too short a time some said—the countess gave birth to a boy, and a full twelve months later to another boy. The two brothers turned out to be very different. One (the older one) was soft, and the other (the younger by a year) was hard. The younger was hairy and large-shouldered although short, and the older was taller, pink-complected, with large feet and a high waist. They showed their differences early on: Imagine, for example, their attitudes toward hunting deer. The younger would, with relish, hunt them for pleasure and the sunday's feasting— the older wanted only to play, listen to the bleat of animal sorrows and relieve them of their ticks. The younger paid no attention to his studies while the older was a favorite of the monks and became the first in the family to read and write—which gave the younger good reason to beat-up on the older throughout their childhood. The countess loved them equally but separately, as befit their separate natures. The count had little patience for the pink one—he felt it overmuch even to acknowledge the little bastard—and when time came for the count to die, he gave the second his lands and title and muttered to the first that he best oversee the studies of some rebellious monks.

And so it went until the invasions by the northern hordes forced the new count to flee into the forest and his brother to sail to Amsterdam where, the story goes, he was warmly welcomed by disciples of the painter Vlad.

The count paused upon this distasteful note, walked slowly past the fast-congealing Walter to the liquor cabinet where he poured himself a large drink. "Help yourself if you like," he said—but Walter did not move, thinking, like the transfixed bird, that salvation lay in stillness. After drinking for awhile, the count began to speak again, this time in a softer voice inflected by the accents of the street. "I'm the last of the line, you know, can't knock Maia up, I don't try hard anymore, rather watch her swing and sway. Maia's not her real name—she took it from some goddess she had read about. I found her in a circus riding horses while standing up; she belonged to the ringmaster who used to whip her after he trained the horses. One night I caught him at it; and after watching for a while, I broke his neck. He looked like a chicken lying there with his head all cock-awry; Maia seemed a bit upset but she went away with me—nothing else for her to do, really. During the war, we kept our distance from the fighting although both sides always gave me jobs to do; but afterwards, I did a favor for a lawyer—the thing I'm good at—and he got me back my castle, without the land of course.

So now we're Count and Countess, and we rent our castle out for plays and such—as royal folks must do these days. Maia likes to act, and I still work for the government, keeping the peace around these parts you know. I also like to help out with the decorations that you see around town; I wasn't much for reading in school but I'm good at making things, and I like it when

I know the story of the thing I'm making. I don't come to this room very often, but I figured you'd show up here—I've been in enough battles to know how the other side thinks.

Oh, stop chattering like that! I'm not going to hurt you; Maia told me you were respectful of her. Anyhow, as I was saying: This room goes back to the olden days; I used to come here to drink and try to find out about my ancestors from those dusty books, but like I said, I'm not much good at reading. You're probably better at it, like Vlad was; we're only half-brothers, you know—and you're a lot like that painter-jerk, I'm sorry to say. But come here and look at what I found. Now get your ass out of that chair, pour yourself a drink and look at these two books! I read in them enough to think that there are some things about Vlad and his father; that should interest you. So stay awhile, read them further, and tell me what is in them about my lineage. We may have both come from the old countess—but our male ancestors were very different. I'm glad of that—not only because it explains the differences in the way we look, but it also gives me a freer hand in how I treat you. So make it your business to find out about the old count, and from there about Dracul himself—that's the one I'm really interested in."

Walter did as he was told. What else was there to do? The count was—Oh—much stronger than he, ruthless, and if not smarter, then more cunning. But he is not an artist as I am, for he prefers real things to artworks—this gives me an advantage—thanks to our different paternal lines. Now that he had expectations of staying alive, Walter spent many days in that room, searching the old tomes and transcribing what he found of interest into the modern vernacular. In the evenings he read what he had written to the count until the whiskey interfered. During these sessions, Maia would sit across the room and knit, looking at Walter from time to time with those indifferent eyes; but she showed no further interest in his talents or his body—for she had known better flagellators and, certainly, fornicators than he. These dissatisfactions became evident on the stage—Maia's yowlings and writhings were increasingly out of sync with Walter's strokes. In one disastrous performance, Walter broke the latex whip and the red dye spilled all over the stage—to the great amusement of a Japanese historical delegation. Director Emanuel did not fire Walter—but took away his whip and demoted him to the soldier's chorus. The pay was not nearly as good, but he had less rehearsal time, and so could devote himself more fully to the archives.

The first story Walter transcribed takes the quest for origins, both Walter's and the count's, back to the era of strife and division, to the time when the last undisputed king, seduced by the perfumed manners that replace events during the dying days of empire, abandoned good governance for his other concerns. In these concerns he was aided by a servant so ambitious, as the story goes, that the man forsook his own virility (he had his balls chopped off) in order to furnish the king with an unsurpassed understanding of the queen's most private being. (He fashioned a cozy nook where the king could watch the queen's many infidelities without her knowing).

The king in this section of the story had a large portrait of himself hung on a high wall in the queen's bedchamber. It was a regal portrait, with a compassionate face and eyes that calmly look across the realm, the portrait of a ruler who had achieved all the good that could be reasonably asked of one with absolute power. The wall on which the portrait hung fronted a secret room which only the king and one other (his loyal castrate) knew about—the workmen who constructed it having long since been sent away to die in foreign wars. And on the wall a small panel had been inserted that opened into the room, and was exactly aligned with that section of the canvas upon which the king's eyes had been painted. Now, these eyes were unusual for portraits of the time; they were painted on separate bits of canvas which could be removed from the back leaving eye-shaped holes that gave an unrestricted view of the queen's bed while they left the viewer undetected.

It was to this room that the king would often go to watch the nocturnal doings of the queen. Oh, he wasn't there to catch anyone in the act—that was another king and an older time. This one had no interest in the recreational cruelty practiced by his forebears; he simply liked to watch what, after a long supportive marriage, the queen and he had tired of performing together. On the few occasions when they, the king and queen, would still make love, they did it in the king's chambers on those nights when she, the queen, would come to return the glove that he, the king, had left behind at afternoon tea. The small connubial delights they gave each other, mostly brought about by well-placed fillips on their various tips, were exemplars of the modesty of their well-regulated kingdom. But given their dimension of regality they needed more, of course—delights more singular and less modest—and the queen deserved it just as much as he, they both agreed, although such radical sentiments could not be voiced in those days, not even to each other.

The other person who knew of the king's secret room—the faithful castrate—was named Vorkur, and he was actually the king's son by virtue of a dalliance, long ago, between the then prince and a Polish chambermaid. Vorkur was a quiet man whose duties as chief purveyor of the king's pleasure had required him to surrender his balls. It was his job, at times of excessive activity in the queen's chambers, to go first into the room, remove the eyes from the portrait, and watch the doings until they warranted summoning the king. It was actually a tedious job, for there were often hours of conversation over many glasses of wine before the queen would decide that it was safe to bare her all to the ministrations of lower others. Vorkur did not cheat, he never extended his vigil into the good parts; at the first signs of seriousness he went to fetch the king, informing him of the turn events would likely take. And then he left the king alone while he kept vigil in the hall.

The queen would often look at the portrait as she played, exposing herself to it more wantonly than she did to her companions—for she, like her king, was more excited by what the world would see if she saw fit to show it, than how she had been brought to that point with no one looking—that is—no one of significance.

Some nights, before her sight would glaze and turn inward with her spasms, she would look at her dear king's portrait and imagine that the eyes rolled and glittered as did hers, and that a muffled moan or two could be heard coming through the wall beyond the bed. But in the morning everything was as it always is, the companions of the night having taken care to carry off all traces of their presence. This king they knew to be more kindly than the last—but nevertheless—fucking a queen is a sure way to lose one's head. Faithful Vorkur, after attending to the exhausted king, watched the clean-up with a smile—and then proceeded to replace the portrait's painted eyes. It is true that at those moments he did permit himself a longer look at the naked sprawling sleeping queen—for he alone in the kingdom had attained the status where he could see her through his monarch's eyes.

The king had other paintings in his chambers: two portraits of the queen, one a smallish portrait head and the other a family scene. On another wall there was a reclining Venus, somewhat pink and pudgy, with reddish body hair and a knowing smile that was directed at an unseen audience somewhere in the painting. On another wall, the larger one, there hung a hunting scene in which a wild Baroque boar plunges through the picture plane in his effort to elude his tormentors. Whether any of these paintings had a room behind it was not recorded, although it might have been appropriate to the queen's concerns that the boar's head have eyes that look at things boars don't usually see in their times of life—like the nocturnal antics of the king. Vorkur would know about this too, for he was a faithful servant to the queen as well as to the king. Yes, Vorkur was ambitious.

Walter was interrupted in this recital of the family history by the lewd noises the count, slumped in his heavy armchair, was beginning to make. Luciano was on the upswing of his drinking and every episode of the see-through portrait took him further from mere listening towards an action that would tend to the pressure building in his groin. The count found it difficult just to relish important things by looking if there were the slightest chance of doing. But, in fairness, while he liked action, he preferred in some instances (more so these days) to not participate. When he killed, for example, he would often prefer that the deed be done by someone else. And he ordinarily had little interest in watching others making love, unless the action was a play of drawn out degredations, broken promises, and tearful capitulations whose consequence was pain.

Nevertheless, after having heard the singular story that Walter had just told, he thought it would be fun if his wife (where is she?) entertained them by playing the queen to his king—a creative adaptation would be acceptable, for Maia had taught him to respect the dramatic arts. After he drank some more, Maia came into the room. The count lurched to strip her of her dress, and managed to pull the fabric to below the swellings of the breasts before she swung her monumental arm and caught him square across the nose—and as an afterthought, she jabbed a knitting needle through his cheek. The count paid these small hurts no mind—after all, he had been in many battles—but Maia's look of loathing made him hesitate. To mask his defeat he turned to Walter and said "You couldn't get it up to fuck her for me—could you?" Walter knew better than to meet Luciano's glare, and murmured softly to the furniture and his manuscript that his is a wimpy little thing fit only for bookish girls.

The three now sat in silence. The count had turned his glare back to the brandy glass; Maia, after wiping off her needle, resumed her knitting; and Walter let his thoughts wander to the vexing concept of watching. When I look into a mirror, he thought, the face I see is not my own. It is the beginning point of the separation between myself and my presence in the world. I twist my jaws and part my lips and my mirrored face smiles or frowns, but I do not know what is meant there anymore than what is meant on other faces that I see before me. Sometimes, to break the barrier between inside and out, I stick my tongue out at my mirrored face, hoping that the sound of laughter could restore the innocence of our presumed identity; but alas no one laughs—my face peers out suspiciously at me, wondering, as I do of it, if I got the joke.

But, as Walter told his face, I do like to watch people I touch get touched by me in films of us. It is a touching beyond sensation. And I also want to hear how my voice sounds in the moment of another time, a time that gives no reason now for me to make that sound. What is it then that is lived in mirrors and in memories? Is it different from what I live right now—the "right now" that is now a second, two, and more and more, ago? The looking glass in this respect is little different from the screen; both images are always in the past and on the outside, yet both seduce with the illusion of a shared present—as if the images of my ingratiating self could not be argued with, dropped from the circle of my intimate friends, cheated, lied to, and in the end destroyed if need be. I accept it; they are free of me, my look-alikes. I can break the glass and burn the film. They let me choose between them as I wish, for I do want to keep in touch. But I don't want to have to recognize them all the time.

I have noticed that the count never looks in mirror; there is a full-length one by the door which he sidles past in the diagonal of a worried crab. True, he is uglier than most, but that is not the reason for the avoidance. The count has successfully—more so than I, although I am smarter—achieved the separation between the in and out. He seeks no praise and accepts no culpability for the actions of the one outside. True, both agree upon the necessity of whatever deed is done, but the relation is indirect, like that between the will and the kicking foot. The count tries to not be disturbed by the outside images, so that he will not be disturbed inside by anxiety—were he to see himself engorged and bloody in the mirror.

Liquor can somewhat alleviate the pangs of an intrusive verisimilitude, but the most hardy liver has its limits. Far better is watching others in the act—particularly when it is so orchestrated by fickle chance or foul design that the players fully believe in the identity of their ins and outs. Torture and rapture are examples that promote this belief—but it's all the same—the fun is in seeing, from further out, how wrong they are.

Walter began to write down these reflections, thinking that he might do something with them at a later date. Maia noticed his withdrawal, and having finished the socks—a birthday present for the daughter of a friend—rose to go, letting the count hold on to her tattered sleeve for balance. Walter had not been able to finish reading them the episode he had prepared for that evening—what with the count's unexpected assault upon his wife—so he called to Maia as they passed through the door that he was ready for another reading whenever they so wished.

You may remember Vorkur—he was the keeper of the peephole of the king. Well, before his formal castration—a requisite for the job, as you know—he had had a son with one of the queen's personal maids who, unwilling to fulfill her further duties—to him or to the queen— returned to her native land of Transylvania. The boy was named Vladimir Draculski, 'Vlad' for short, in honor of the Great Impaler, the founding father of the empire; 'Dra-cul' made reference to the preferred site for impalation; and 'ski' meant "son-of."

The weight of such an illustrious name made it difficult for Vlad to participate in the, often brutal, games that the younger men in his circle were expected to play. Oh, Vlad had all the will in the world, but he had a habit—no doubt inherited from his father Verkur—of watching himself while in the act. and so he would often disconnect—which we all know is not good for authority or erections. Nevertheless, as he grew he came to enjoy the uncertainty of performance as well as the watching. Together, they gave rise to other sorts of practices, and because of these admittedly eccentric preferences, he soon gained the reputation of being someone who is quite accomplished in distaff ways of making love. So he seldom lacked for companionship, and his partners did not seem disappointed; for it may well have been a relief for them to join in his search for imported novelties, primarily French and Italian, instead of making do with that same old Balkan thing—the good old in-and-out and bless me father I have sinned—practiced by the other boys and girls.

Vlad lived in a fortunate time, when the court had gained some independence from the clergy, and the holy prohibitions on food, sex, and reading matter were reserved for the high holy days and did not seriously interfere with the education of the sons and daughters of the better classes. Despite his given name, Vlad came from servant stock—but he quickly saw (an innate talent) that some services are higher than others, and if you understand their heirarchy, it can be used like a ladder pulled up behind you as it's climbed, and eventually you find yourself consorting with the children of the well-placed folks your parents had waited on. Sounds crass it does, self-seeking too. But what else was there for Vlad to do except become a priest or soldier? The first leads to celibacy or worse and the second to an early death—not much of a choice. His friends agreed that Vlad was just being sensible about his life.

However that may be, Vlad became an artist, for he saw that artists have an easy entree into the chambers of the rich and powerful, and are usually excused for their pecadillos and other bad behavior because all know that these are earmarks of a creative temperament. At first, Vlad specialized in portraits, for he found that the sitter, if not the subject, was usually more interesting and well-connected than the pastoral alternative of cows and meadows. Although he was naturally adept at capturing likenesses, Vlad also had historical ambitions and found he needed more sophisticated painterly skills; he practiced long and hard to master the flow of draperies and the translucencies of lace as well as the latest innovations in perspective. Soon he began to receive commissions for full length portraits, mostly of ladies—for the men would not sit still that long and were usually away at war or hunting. Vlad was making good money by this time; he took to wearing velvet jackets when he painted, and fine Italian leather boots. His manner also became more polished, and he collected spicy stories with which to amuse his sitters while he combined his formulaic renderings of skin and raiment with the specifics that would identify each sitter's portrait as her own.

One early afternoon, the Countess Luciano came for a first consultation. This was an auspicious commission, for the count was quite enamoured of his young wife's radiant looks and would pay a pretty penny to have her smiling down, secure behind the paint, upon his admiring friends and jealous enemies. Vlad chatted up the countess with enthusiasm, extolling her aesthetic charms and telling her of the great portraits of the masters wherein noble ladies were usually portrayed allegorically—the older ones as virtues, as chastity or charity, say, and the younger ones as goddesses, the ones athletically inclined as Diana, and those of a more leisurely and introspective cast as Aphrodite, the goddess of love better known as Venus. The countess was much intrigued, for she had a taste for scholarship, and asked animated questions about these portraits: How many sittings were involved, and how the allegory and the pose most suited to the subject was identified. Vlad was circumspect, situating his answers within current Neo-Platonic discourse on beauty and the sublime. When the countess grew somewhat impatient and became figity and uncomfortably specific, Vlad retreated to his archives and brought out engravings after famous Venuses—which he had once bought when on a trip to Rome—and laid them out before the countess—who now insisted that he call her "Maia." This name, as she told him, is not her given one—"Maya"—which she now dislikes, but which, despite her wishes, continues to be used by the elders of the court—who are quite doctrinaire about matters of lineage. One reason, then, for the portrait we are contemplating, is that her preferred name would be prominently displayed in the title—"Countess Maia Luciano"—as a corrective to those who, for their silly reasons, prefer the old usage.

The countess, priding herself on her learning, explained her preference to Vlad in some detail: "Maya," she said, is actually more a description than a name. Its origin is in the ancient Indian holybooks, the Vedantas, where it is used to counter the illusion that reality is to be found in awareness of the self rather than in Cosmic Unity. The term is so used in the phrase "The Veil of Maya" which identifies the distortions the individual self gives to reality. But then the term travelled to Greece (we all make trips, don't we) where it names the daughter of a Titan who was impregnated by Zeus (a nice way to demarcate the passing of dynasties). She bore him a son, the god Hermes, and somewhere in there the name changed from Maya to Maia. (As I, myself, am not an illusion, and as I adore Zeus, I prefer the Greek usage to the Indian).

There is still another variant—"Maja"—used by the Spanish to designates a belle—a beauty who transcends her origins, as I have mine, by virtue of her own talents and endowments. While I think that Maja also fits me well, it remains a social designation, applicable to every pretty and privileged young lady in town. I need something more—a name which is uniquely mine, but which carries with it intimations of the other spellings. The Titans daughter, the one ravished by Zeus is Maia—that was her name, as it will be mine—when I get it past the obtuseness of the court.

Maia then lifted her left arm to reveal the luxurious hair that adorned her armpit and said to Vlad: "let's look at those Italian drawings of goddesses, both the nude and naked ones."

In the time it took to look at the drawings, each minute stretched endlessly into the days to come—and from the corner of the room Vlad could hear his voice extolling the qualities of the various prints: the Botticelli and the seashell's foam, the shimmering skin of the Tintoretto, and especially the Titian—ah, the incomparable Venus of Urbino. Maia looked intently but was silent; she was not obliged to make conversation, only he—but he too fell silent, busying himself with tidying the scattered piles of prints and worrying that he had gone too far. Between that meeting and the next, Vlad could do no other work; he was now sure that the countess had seen the prints as pornographic (indeed, he had shown her prints by lesser artists which depicted divine and other forms of coupling in great diversity and detail) and so Vlad could only sit and sweat and wait for the sound of soldiers' feet. A soldier did come, but only to give him a note saying that the countess wanted the next meeting a few days early—and would he have some studies ready? Furiously, Vlad began to draw the goddesses he was most familiar with in all imaginable attitudes and disportings; he borrowed shamelessly from every source he could remember, and his final drawings, made just before Maia arrived, were so focused as to portray only those aspects which exemplify the beauty of the subject—all the rest, everything extraneous—even armpit hair—was eliminated.

Of course, he did not show her all the drawings, not until much later, not 'til shortly before the count ordered him beheaded. So they talked abut the Titian—Vlad and Maia did—about how regal nudity can be, how nudity and nakedness were really very far apart—how every Venus, whoever the painter, seems poised in expectation of her absent lover, possibly someone beneath her station—for was it not Venus who early in life had grown weary of the crowd on Mount Olympus, and went searching for—what's his name—Adonis?

A few more sessions passed—Vlad was not charging by the hour, and the count was fighting a fairly distant war—before Maya brought up what had been on her mind. She thought, she said, that she herself was not an introspective type—and too young to be depicted as a virtue; so if it were to happen that she should be painted allegorically, why then it could only be as a goddess, as, say, Venus—"don't you think so Vlad?" "Oh my yes, you would make a wonderful Venus, radiant, more beautiful than the lady from Urbino, and I only wish that the great Titian were alive to do you." "Vlad, tell me, you know the artworld well—I have been a maya, and am already on the way to being Maia, but is there anyone around who has the skill and strength of character to make me Venus?" Vlad thought a while before answering, for his head was full of her dangerous perfume: "Well, there is this young man in Rome, very good painter but rather nasty, and it is said that he prefers to paint boys; then there is an older Flemish artist, well regarded but expensive; his nudes tend to be fat, and anyway, I hear he is busy painting portraits of his new young wife—besides, I think we will soon be at war with them. The only other one in that league is a reclusive Dutchman, who seldom paints the paintings people want, is more enamored of light than flesh, paints himself a lot, and loves his wife who is a rather plain and dumpy person adept at managing his affairs. But, if you like, I'll get in touch with them and try to get them here so that you can choose, for they are all quite good." Maia fidgited a bit and slowly stretched, arranging her limbs into a Venus-pose. She then sighed as if with boredom, and said all that was a lot of trouble, foreigners are always undependable and expensive, and there may be a good solution right at hand. She stretched some more, then looked at him directly: "You will paint me, Vlad, yes you. But I want two paintings, not one. You will make one painting for the count—it's his money after all—and one for me. The first will hang in the great hall for all the court to see, but the second will be hidden in my private chambers, behind the curtain over my bed, where none dare go but those whom I invite. The paintings can be very much in the same pose, but in the one I shall wear the finest robes and lace, and in the other I shall be nude—as close to nakedness as your talents allow." And then she laughed, that tinkling laugh reserved for ladies of the rank of countess or above.

So the tandem portraits began. Maia would first pose in all her finery, but before an hour had passed she began to strip—first the heavy velvets and the patterned boots, then the frilly damasks and lacy underthings; she would throw them together in a heap, stretch her arms above her head, and settle back upon the couch, her eyes moving between Vlad's face and the details of her naked body. In truth, Vlad painted better when her clothes were on, for her body was somewhat disappointing, with the breasts quite far apart and the legs too short between the knee and ankle; also he shook uncontrollably each time she disrobed. But the paintings had to move along—it was a commission, after all—so between sessions Vlad hired a young serving maid to emulate the pose. She was quite honored, this pretty red-haired child, for she had never expected to be the object of an artist's eye, and she appreciated the evident delight Vlad took in her—it was a good thing too, for later she saved his life.

The count returned about the time the portraits were completed, and before long he found out about the naked double of the official one—for in castles there are no secrets that can long be kept. A beheading was prepared. But before the soldiers came for him, Vlad was warned by the red-haired serving maid—whose boy-friend was a soldier and just had to tell her the hilarious story before the painter was arrested. Vlad thanked her for her warning, gave her half his money—to care for the child she was now bearing—and kissed her upon the breast and foot, as he had often done. He then gathered his best brushes, some winter clothes, and made his furtive way from the castle to the harbor, where he found a boat that for half of his remaining money would take him to Amsterdam, a city known to be hospitable to strangers and hostile to extradition.

You might wonder what happened to the countess Maia. Well, nothing really. She insisted that the naked portrait was not of her: "My dear Luciano, you can see that the breasts and especially the legs are all wrong—common, really. That dastardly painter must have made it up; just as well he's gone, and good you didn't pay him—but we needn't throw the painting out; let's hang it in my chambers where it will inspire you each time you come to visit."

In Amsterdam Vlad found a modest room and took to wearing simple workmans' clothing.

He grew a beard and made his living by drawing charcoal portraits in the town square on weekends. He was determined to regain his life, avoid women, and become a great painter. At night he dreamt of the Flemish masters, their last judgments and annunciations, as well as nudes of a different kind, like Georgione's, that co-habit with the landscape. He worked methodically every day, exploring the great themes and trying to capture the grandeur of the older styles. He didn't show his paintings much to others; he had made few friends in his new home, and he also felt too set in his ideas to solicit criticism from the other artists—what did they know, after all? So he painted on, accumulating what is called "a body of work" which, someday, he would present to the world for inclusion into history.

But it didn't come out that way. Despite the purity he had brought to his paintings and his new life, Vlad slowly came to realize that, notwithstanding all his efforts and abstentions, his talent remained a modest one. He hung his finished paintings around the room in which he slept and worked—a chronology of his journey; and he looked at them in sequence everyday, hoping to find the larger realizations that he knew were in him. But the paintings went the other way; they became less and less what he had seen in them when they first were finished. Why could he not have pulled them further towards that vision, the one he sometimes caught a glimpse of—just after a particularly daring stroke? To be sure, his painterly subjects were ambitious: noble crucifictions and bloody flagellations, wild Venuses in the woods, and even some bowls of fruit to show his formal skills. But after they had hung around awhile, they just looked pleasant, emulating the good manners Vlad had so assiduously learned in the castle, back across the water. Someone—he didn't remember who—once said it is a difficult moment when you realize that the greatness you took for granted as an only child is not yours after all.

Vlad hid his depression well; others thought of him as cheerful if aloof—but he could not rid himself of disappointment with his transcendental failure. He knew, of course, it was not his fault, not being great—just a cosmic lottery. But still, he had been close to greatness, close enough to glimpse what it might be like—that region beyond the heavens where the forms perform the minuet within the confines of austere verticals, supportive horizontals, and subtle diagonals—and he could not easily put aside his early straying into decorative boudoirs—a shortfall that will shadow over his remaining life.

Vlad loved himself too much to think of suicide, so he painted less, and took to finding ways that would fill his resultant emptiness with worthwhile and exciting things. Spring was coming; the air was clearing; the braided girls strolled on the friendly dikes in the sunshine of the early afternoon—and the clacking of their wooden shoes did not sound so silly after all. He forgave them for their simplicity, and he forgave the others too. After all, the countess was not to blame for his troubles; she just behaved as countesses do. And he should not blame himself; he behaved as his father had advised him, namely, that servants' sons should look to women for their betterment. And the cosmos, after all, is blameless by its very nature—it just is. Basking in the unexpected glow of this conceptual leap, Vlad thought that the future was beginning to look promising again.

Having recited this section of the manuscript, Walter fell silent. The count had drunk too much to comment or indeed to move, and the only sound beside his troubled breathing was the click-clack of Maia's knitting needles as she made her way through another multi-colored sock. Walter wished that he, like his ancestor Vlad, had gone to Amsterdam, that in his peril he had fled straight to the harbor rather than to the recesses of the tower—there to lose his will again and become captive archivist to the family Dracul. But had he escaped and crossed the waters, it would have been to another Amsterdam—not the one in which Vlad had found his soul and lost his art.

Walter remembered hearing from a poet-in-residence that these days Amsterdam is filled with placid women sitting in store-fronts selling well-chewed bubblegum. And then there is all that pot, they sell it on the streets, the very center of the plant, the fire of forgiveness that helps to not discriminate between what one does or not.

Walter, you must try it out. Bring Vlad up to date why don't you, and imagine joining him in Amsterdam today. While you write into the night he paints through the morning when the light is good. Then, burdened by a too clear head, he takes a poke or two to make the freshly painted passages sit in harmony with the rest. Having satisfied his art, he walks with you into the slanting shadows of the afternoon, slowly—a third poke at the fragrant flower ensuring that there is no need for haste—to find some friends and beer that will cut the dryness in the mouth. There is talk of art, more or less depending on the season and the shows that come to town from faster places; and there is talk of store-front girls and whether they out-class the ones who walk the streets.

It is your job, Walter, to sort this out; you are the one who has the way with words. "Very well," Vlad would say to me, "this is the old classic-romantic problem, isn't it? Must we stay or can we leave? The placid ladies seen through the window are pure form—no movement to entice you and no expressiveness to sway your judgment. Each woman finds the pose that is her ideal and holds it through the afternoon, breaking only at the opening of the door—whereupon she barters for her moving essence in the privacy behind the curtain. But that's the trouble with classicism; you never know what you've got when you only see it through a glass. It's all a pious show, the stillness and the measured space. Because the forms don't move you think they won't outstrip you, leave you to your failures—still convinced that you can have it all until the morning comes. Yes; slow sex, slow steady silent sex, the kind that brings you home even when it's time to go. There is no death, the poet said, in such a show."

But now let's give the other side a chance. Should we not prefer the transient and ephemeral art, the circulating hips across the cobblestones and up the stairs and out into the streets again—slowing just enough for a transaction and some whispered words about how big and bold you are? Leaving is romantic, even when it outstrips memory and violates the picture-plane. To tumble down the moving flesh is a fall across the middle of your life, a journey that ignores the dying waiting at the bottom—down the line Ho Ho you go, past the dirty feet and pretty painted toenails.

Vlad is lucky he became an exile in old Amsterdam, the locus of ground lenses and the geometric method, the origins of democratic light on a face which turns and looks at letters on a tablecloth. It is a place to fail without distraction and then walk into the shadows without regret—a fleeting diagonal across the upright doorways of the afternoon.

In truth, Walter would be better off could he have joined Vlad there and then; but Vlad didn't offer, left no missive of instruction for his descendents. As it is, Walter now has to read through all those dusty books, and tolerate Maia's pointed indifference, to learn what little could be known about the time of Vlad. But large gaps remain between the then and now, too large for anyone to venture back to there from here. So the time has come for a terrible bravery, one which Walter—because he is such a coward in the ways that count for others—alone can muster. It is a bravery of pure narration, time to find beginnings and make them responsible for what has passed thus far.

Beginnings, much like ends, seldom lie along straight lines. Those who have practice in searching them out, often spot them (there are always more than one, you know) cavorting with wayward middles in the fields behind the bushes that veil reality. But interesting beginnings can also be found in musty tomes—such as the one that Walter has been reading in his efforts to mollify the count, and to find a more plausible linkage between himself and Vlad Dracul.

What follows is a passage that he found in an appendix to a commentary on a particularly obscure section of the text. Neither the writers nor the authors could be identified.

Once, between the early and the later times of writing, there was a village in the lowlands where people who belong together lived together and apart from others. Their numbers made them more than a tribe but less than a nation, and where they lived is now a distant place greatly altered by the movement of tectonic plates. What they called themselves is not recorded, but we know they were survivors of an ancient journey whose perils weeded out the meek and weak among them, so that when they reached their haven, they were spare enough to begin a new and hardy race. They spoke a number of dialects all of which had roots in a common language, but the differences identified their separate families and permitted them to distinguish between the deepest secrets they would share only with their loved ones, and the broader knowledge which brought them together as one people. They had nothing but disdain for the indigenous tribes who were there before them, and did their best to discourage these from any further proliferation; and while they usually hid from large bands of strangers, they could be quite hostile to single passers-by, robbing them and using their skin for drum-heads.

During the fifth decade of their settlement, a young man named Gavagus was born. He was so called because like his namesake-rabbit, he was fleet of foot and could run circles around his adversaries, and outdistance all the emissaries sent to convince him that it is good to suffer in times of drought and famine. He wasn't up and running all the time, however, for he was ambitious, bright, with a self-developed interest in success. His problem (as we put it these days) was a socio-sexual one: a premature conclusion when something—most any thing—was just beginning. Perhaps, at first, he didn't think the things in question deserved more than the time he gave; later he felt that he would never fail if he could always leave before things ended; perhaps he was just too afraid of the dark to stay.

When Gavagus was young, most thought him charming, and there was quite a merry game of seeing who could run him down and keep him working for awhile. It usually took two to do it, so including him it came to three, not quite two pairs—which allowed for fickle matings of success and failure with joy and rage (which does not count as four). Because the others ran after Gavagus a lot, he bypassed success and came to rage—that insistent child of failure—quite early. But he continued running until someone caught him who had excess joy to spare. Young folks who give away free joy think that, like loaves and fishes, what they give will multiply. Not with takers like Gavagus around, for such as he will eat and drink and love and leave (which should make four). Such people never offer to clean up, nor help in planting the watermelon seeds; they do not give the dog a porkchop bone, and will not join the friendly farting that closes a communal meal (definitely four).

As Gavagus grew older, givers grew more wary of his appetites, and he began to show the wear and tear of fending for himself. It was not sagging skin or wattles that he showed, but a look of distraction that molded his face with creases around the nose and pouches beneath the eyes. The flitting expression of reproach which he turned off and on when younger, now settled into place; it became the permanent feature of his every shift and change. In time, he came to look as if he never is completely where he is. Notice that he does not hold your eyes while talking but looks outside the window and checks the distance to the door. Watch closely—for he is about to move: The gesture last made while talking extends and pulls him to his feet, then in the sudden lull a quick nod at the hostess, and he's out the door. You can see his skinny form quick quickly bounding down the garden path, over the bushes then quite gone.

It's not that Gavagus was uncaring, it's just that no-one else was ever quite as real to him as he; and as long as he stayed within himself, where he went was as much here as there as anywhere. But he had to eat—so he took up a foreign custom (one then much in vogue) and began to make himself into a work of art—to make himself out of himself, so as to be both nicely balanced and self-sufficient. After a time he came to like his artness well enough to think there would be no problem getting success, joy, and food. Well, yes and no. He did not starve, but found that increasingly he was being offered plainer fare—the cognoscenti said he showed too much classicism within his art to satisfy changing tastes. What is left for Gavagus but rage?

To be sure, Gavagus never as a child had eaten from a silver spoon; he often ate fallen apples with large worms, in the pouring rain, deep in the primeval forest. But when he grew and became most beautiful, he was offered rabbit braised in red wine sauce with pickled matron for dessert. As he wanted to keep this going, he needed something more than some well-shaped limbs or an agile tongue. So he gave them art, smart-art, Gavagus art, festooned with true belief and inferentially linked to deeply held religious and political ideals—good value. So why was the food getting worse?

Nothing was getting better. Gavagus had been mistaken about the nature of the inferential chain; he thought the links could be fashioned in his mind and memory whereas the true chain is found only on the hard ground, stretching on busy roadways into the center of the town, there to pass through the portals where the high and tasteful live. When Gavagus finally heard about the true chain (a tardiness we can attribute to his incessant running) he went out to hook into the link he thought was his. But the older ones who barter food and joy for art and rage, deceived him. These sages know all about chains and where the real one is; yet part of their amusement—indeed, a requisite for their high office—is to strew about all sorts of chains and let the wannabees, like Gavagus, hook into the loose and barren ends. These hopefuls are then encouraged to give their chosen chains a yank and so interpret the small resistance as success—an occasion for a little joy—until the unexpected appearance of an unconnected end coming through the studio door, gives rise to a lot of rage.

Rage is bad for running; it carries the weight of another's provocation and slows you down. So Gavagus, why do you run? Ease up! Stop, in fact! Forget the chain; put some ketchup on the food you have, eat a lot, and offer someone else a bite. Yes, stay around and become a dispenser of free joy—even at your age. Many will be grateful and sing and dance around you, others will smile and talk with you if nothing else. And don't mind that there are some who take the food and leave without so much as an "umm-it's good"—you know how that is.

Walter was so moved by what he had thus far learned of Gavagus that he diverted his scheduled reading into a writing, the first in many months, a rumination on his new life now that it could look back upon the old. So he gave his audience, on that cool September evening, an unexpected commentary about the perils of finding a self by learning how one's ancestors really are. The count had been nodding for some time, but finally sensed that the story was not unfolding in the direction he liked—he much preferred titillation to sermon. So he clenched his massive hand and fixed Walter with a bloodshot stare. But before he could complete the gesture—one he had used many times—he slumped back in his chair, almost but not quite overturning the brandy bottle. "Walter," he said, "you're not a bad scut, but you're full of shit, you know. I want you to tell me about Dracul, not some wispy country artist—why is this fucking family so full of artists? And no more snide remarks about the fashionable folks! I happen to like those skinny models, all pointy elbows and big bony knees—it's something different, a change from all that baggy flab." Walter had always tried to imagine how couples he met would look in bed. So he overlaid the count with a twiggy-type and tried them out in various tops and bottoms and a few inversions. He surprised himself with these images; because they lacked a middle-ground they were quite abstract, a unity of opposites, a harmony of incongruity, the coupling of a mastadon and a spider monkey. The count then rose to his feet, and with deliberate although unsteady steps went to relieve himself for the night. As he passed through the door, he gave Maia the furtive and apologetic look with which, especially in the recent democratic years, he would begin his attempts at intimacy.

Maia let him pass and looked at Walter with eyes in which a small new interest now appeared. She had not thought, before his latest reading, that he possessed such an asymmetric soul, that his layers were not like the rings of an onion, the inner simply replicas of the outer ones; instead, beneath the layer he had reached this evening, she saw the inversion of much that earlier had presented itself outside. Peeled closer to its center, Walter's soul became rotten: It changed in form and color—its elements lost the pudgy pallid pink of their exterior and became an acid green and gray, encrusted with the rage he had just attributed to Gavagus. His internal intentions were suffused with prickly envy; and his covert ideals, those mainstays of his academic days, were clouded over by a fulminating disappointment. How nice; how sexy.

There is evidently more to him, Maia thought, than she had noticed in the dusty sticky afternoons upon the hill. The lineage of a beast is written in his wrinkles; yes, and the ancient hatreds glint beneath the candor in his soft brown eyes. Lust—even love—with such a one is worth pursuing; but there was more at stake for her than pain and pleasure. Maia for the first time recognized that within her fleshly fantasies there sat an obligation. She slowly put away the sock she had been knitting and rose to follow the count to bed. As she passed Walter, she touched his shoulder and said "I want to have a child with you."

For a reason that were not clear, Walter felt no surprise at Maia's statement. He remembered only that it had been made before, long ago, by someone they both knew; and what they knew had come about because of Gavagus—not the one so far narrated but the one he would change into. Walter had by now lost all fear of the count: Luciano's drunken incomprehension had shrivelled him to the stature of a pensioner, his head only a pea embedded in his massive shoulders. Because of his impotence, the count was now a discard from his own history and could no longer influence the story being told—in which Walter would impregnate Maia with full and fitting ritual upon the mountains, the forbidden region to which Gavagus had once fled. It remained for Walter to learn the customs of the ritual, how much is owed to the original intent and how much can be improvised to fit the present need. For this, Walter had to subject the chronicle of Gavagus to an inquiry that would be more critical than before. So, with the silence of the castle as his sustenance, Walter read on.

Unlike the descendent who later took his name, Gavagus in his peril had chosen to go up instead of down—to the mountains rather than the ocean—and so he renounced the transparent light and shadows of the lowlands for the penetrating cold and indifferent savagery of the upper region, a choice that made him almost die before he became a king. He reached a small village, but when the elders of the village denied him the food being eaten off the communal carrion—and then threatened him with pointy sticks—Gavagus ran with a speed he had not achieved even in his younger days. After many days, he reached another, larger, perhaps more hospitable village. True, no one threatened him, and he was free to rummage through the discards. He approached the matrons of this tribe, hoping as before, that they could become a faithful clientele. But this did not happen; the women had their pick of tractable and tumescent apprentices from the local school of art, and did not want to consider this skinny unkempt outsider. Gavagus then played it straight and applied to the bureau of cultural affairs for a commission—he told them he was good at painting street scenes as well as landscapes and could imbue either with multiple meanings. But the officials told him that they only support local artists. As Gavagus did not wish to go skinny into his dark night—not yet, anyhow—he ran, more slowly this time, hoping to catch a squirrel, or find some fruit-trees.

But the summer was now ending, and the cold night air joined with the bugs in biting at his exposed flesh. Gavagus had only a cloak, a knife, thin leather shoes, some crackers and a flask of brandy—these latter having been snatched from the last vendor at the edge of the village. The imprecations following him faded with the last traces of the path, and he now climbed slowly onto the hard rocks, his direction pointing upward and away from the fires that marked the limits of the village below. His shoes soon gave in to the rocks, and as he knew he could not go far with bloody feet, he carefully replaced the soles with tree-bark, stuffed the uppers with leaves, and tied them to his feet with vines. Although, as a child, he had been taught the usual lore of hunting and setting snares, he subsisted for a time on plants and grubs, washing down each swallow of a stubborn root with a sip of brandy. The grubs were less aesthetic than the plants but safer; even so, he twice mischose his plants—and the periods of retching left him dangerously weak.

When the snows came and the ground froze, there was little edible to be found, and he began to feel the onset of his death. To have his clean-picked bones adorn a gulley near the wild high peaks is not so bad an end, he thought, if only someone far away across the seas would somehow learn about it, and then give his death a story. But he knew this would not happen, not with his desire to live, so he continued to move, thinking that motion held off death through its changes—and who knows, change may be for the better. And he was right. He scared a small bear away from its kill, a young doe, and he ate the flesh, first raw then cooked, and used the hide as shelter and the bones and ligaments as tools and weapons. So he put aside his death and began to emulate the mountains, jutting and inclining with them as the elements dictated, and he made no plans beyond choosing a good traverse and following the signs of game. He lost all sympathy for the animals he killed, and thought of humans only as his enemies.

So when Gavagus did meet another human, a small man plodding through the snow—a scribe journeying to a remote village who had lost his way—he immediately attacked and was about to cut the small man's throat when he heard the sound of language—a plea to spare a life and a promise of servitude. That he understood the sounds as words surprised Gavagus, for he had not thought in words for some time; but the sense of what he heard was agreeable; he had more to gain from his victim's servitude than his death—and he could always kill him as he pleased.

Gavagus now had a slave, a first member of his own kingdom, and he was now free to hunt throughout the day, knowing that the cooking and the sewing and the fashioning of tools would be done for him. His slave was named Vorkur, and he gradually began to refer to himself as a servant, which Gavagus accepted—for he did work hard and was good at managing affairs.

As the seasons changed, and the earth of the paths began showing through the snow, they met others in the mountains. Each time Gavagus leapt to kill them, and each time he desisted, swayed by a plea for mercy and a promise of eternal servitude. Although Vorkur took no part in these attacks because he was weak and unskilled, he was also wily, and with each newcomer to the group he took it upon himself—Gavagus not being interested in details—to apportion out the chores and food. So it turned out that gradually—very gradually, for he remained afraid—Vorkur became the administrator of what had now become a tribe; he divided duties and loyalties, set up checks and balances by establishing a system of informers, and so ensured that things would run as Gavagus would wish.

Vorkur knew, for example, that Gavagus had a taste for the show of pain—perhaps intrinsic perhaps acquired, it didn't matter—so periodically Vorkur would sacrifice a member of the tribe through an accusation of disloyalty, and then would entertain his master with a pageant of torment which elicited peals of laughter in grand cacaphony with the shrieks that rose and fell throughout the evening. Gavagus's laughter was so potent an expression that it was imitated by others; and soon the raucus laugh, starting high and ending low, became a form of greeting between the members of the tribe. The popularity of these and other pageants fostered Vorkur's rise in power and esteem; but he always took care to please his master, and concentrated on ways to expand the tribe's domain.

Eventually, their numbers increased and they—both men and women—were molded into a fighting force. Discipline became a way of life, and at one point, Vorkur thought them strong enough for him to propose that they raid the village from which Gavagus had once fled. There was much wealth and many hostages to be had, but most of all there were buildings, farms and animals, all waiting for a ruler who knows his mind.

So it happened that Gavagus and his horde swooped down the mountain one dark night, and through the ancient tactics of murder rape and pillage, conquered the village before the morning light. Now the highlands and the lowlands both belonged to Gavagus, and the number of his subjects—for he was selective in his killing—had much increased. It became evident that the resources talents and technologies at his disposal demanded that his identity as a nomadic mountain ruffian be changed to something larger, grander, more appropriate. It was Vorkur, of course, who came up with the bright idea: "You should be the king of this vast domain," he said, "a king with ministers, an army, tax collectors, a whole host of courtiers to entertain you and serfs to contribute to the infrastructure. We'll have a magnificent coronation, find a regal name for you—'Gavagus' has too many primitive associations—and you'll do the kingly thing of delving inside for the appetites that you now can only satisfy on the outside. All this will be accompanied by pomp and merriment supplied around the clock by your faithful subjects and a well-trained staff—I'll see to that. Of course, those who disobey will be met with immediate punishment, justified not only by concerns for state security, but by your status as absolute monarch."

Things worked out pretty much as Vorkur had planned. Gavagus—they had not yet found another name—spent the first few months taking his revenge upon those in the town who had previously slighted him. It took that long because his memory was refreshed with the information given by each successive victim, and also because he had decided to circle the boundaries of his domain with their impaled corpses—an idea he got when he saw a hunted animal being transported in that fashion and thought it an improvement over the usual slinging by the legs. He liked the effect so much, he kept the killing going—which was not hard, as there were many suspects in his newly conquered domain.

With his cares diminishing and his needs well catered to, Gavagus began to indulge his once thwarted appetites for food and drink. He no longer ran, got fat, and spent most evenings in an alcoholic daze. The drinking further loosened his grip on the affairs of state but it also increased his paranoia, and he spent the sobriety of his mornings concocting plans to protect his power. He summoned the most brutal and effective of his fighters, a short man with huge shoulders and a small head, and directed him to assemble others who were most like him. He then picked a score of them, gave them money and swore them to his personal service; he warned them to take no orders from any other, to report all things suspicious, and kill upon command.

Vorkur did not much like this turn although he understood; but Gavagus noticed his reticence, and one festive evening lunged at Vorkur with a sword, diverting its direction at the last moment so that it skewered a servant holding a tray of fruits and nuts. Gavagus bit into an apple, looked Vorkur in the eye, and laughed his high-low laugh which was immediately taken up by all those present. It was then that Vorkur sent his only son, also so named, to a more civilized western land where he would be safe.

When Vorkur dared approach his king again, it was with a list of names that would be fitting nominations for the royal person. He had chosen them carefully, linking each name with a distinguished personage or exemplary quality, and supplying each with a supportive historical context. He picked a thursday morning for his presentation, a middle-of-the-week time, a morning when Gavagus enjoyed a relatively low level of hangover and suspicion. The list of names had many worthy of consideration—Vorkur, after all, was monastery trained—and Gavagus had great fun intoning them while viewing himself in a mirror. "'King Maximus I,' certainly up there but a bit impious—no need to upset the gods unnecessarily. And where did you get these? 'Rictus the fair' and 'Judas the Just;' names my tax collectors would use when they shake down the lower classes—anyhow, let's skip all those with 'us' endings, too foreign; I want something from here-abouts. 'Cretinian the wise' isn't bad—but a little too far south. What about this one, 'Dracul' ?" Ah, this was Vorkur's favorite, for it had historic and mythic, as well as descriptive qualities, but he presented it in a neutral technical way. "The name combines 'dra' which signifies a going towards, a quest, a mission, a penetration—with 'cul' which has the meanings of center, essence, end.

You see, master, your virtues, ambitions, goals, and pleasures are all included in that name. Its etymology reaches back to the first conquerers of this land, a fierce people who would rather drink the blood than eat the meat of those they killed, and so, as the legend has it, they grew strong and lived long—although it is also rumored that they were quite sensitive to daylight and had trouble adjusting to solid food when the blood ran out. But, on balance, I think it a splendid name for a king. For your first name, I chose 'Vladimir,' 'Vlad' for short. The Russian flavor gives us greater range—a suggestion that we have powerful friends abroad—and the informality of 'King Vlad' will endear you to your subjects' hearts. What do you think, sire?"

Gavagus said nothing, but the next morning he summoned Vorkur with the message that Dracul the king would speak with him. Vorkur bowed, smiled, intoned the name, and bowed again; he then pulled out a schedule for the coronation festivities: "Non-stop dancing in the village square, outdoor roasts of pigs and oxen, all restrictions on public drunkedness and lewd behavior suspended for a week, and—as it is important, sire, for the populace at first to love the king—a gold piece for the head of every household. I also have assembled three guest lists, the most inclusive covers the village-wide festivities you will watch from your elevated throne; then there is a smaller list of notables permitted to approach you and swear their eternal allegience when they catch your eye; finally a group of intimates who will join you in the private pleasures that grace your elevation. This last list is critical, for its members should function as your royal family until you get around to making one. I will watch them carefully as they carouse and vie to please you—I, myself, am always sober as you know—for it is on such an occasion that some few might reveal an undue forwardness or a hint of disrespect foretelling a later disloyalty." Vorkur paused, for he had come to the delicate part: "Your majesty should also think of awarding rank—barons counts earls and the like—a royal court reinforces the stability of the royal house. As for me, sire, I would like to be known as your minister, preferably with the prefix 'prime' attached; this will give me authority to carry out your wishes and confound your enemies." Gavagus—now firmly Dracul—threw a goblet of morning wine at Vorkur's head, but not too hard, so Vorkur knew that he could proceed with the organization of the kingdom.

The coronation exceeded all expectations. Brave men died before their time, scores of children were conceived, livers were strained beyond capacity by the flood of drink, and all revelled in the sprinklings of sweat, the mingling of flatulence and perfume, and the engorging of the communal gut. Vorkur had spread the message that under the new king the future would remain much like the present, and from the viewpoint of a bacchanale, this seemed a most agreeable prospect. How Dracul celebrated the later hours of his coronation was never recorded. During the following weeks, six on his most intimate list of thirty-one were impaled and distributed about the outer gates; and the rest, very conscious of the noble rank they had achieved, nodded wisely, and never breathed a word of protest to anyone.

So it was done, and Drakul turned his attention to the duties of a monarch. He sent his troops, under the command of the man with massive shoulders—now made a count, the count Luciano (named after a premonition of a line of so-named counts) to range beyond the informal borders of his kingdom, kill all they met en route, and extend the signposts of his rule up to where they saw traces of a hostile army and another kingdom. Vorkur thought it premature to engage in a protracted war, so he turned the natural bellicosity of his king in the direction of making heirs to the throne, another necessary duty. Confronted with the array of women Vorkur brought for his inspection, Dracul showed little enthusiasm beyond an occasional quickie to uphold his sobriquet of 'Vlad the Impaler.' Faced with the daily parade of nubile hopefuls, he thought of his time as Gavagus and the sexual hoops the up-scale women of the village—all now dead, of course—had jumped him through; and how his difficulty in showing enthusiasm of the kind that cannot be feigned, marked the beginning of his loss of favor and led to his eventual flight. Well, all to the good, for he was now a king—but he did need heirs; and this proved difficult because, in truth, he could only become aroused by fear and pain, and even so he preferred to watch rather than to disturb the poignant sounds and jerky movements with his half-hearted interjection.

One afternoon, Vorkur brought him a woman named Maya, a foreign woman from the west, named after the word for "illusion" because she had been drugged and kept well out of sorts throughout the abduction that brought her to this land. However, even in her delirium she showed a will to survive—so she would enthrall her various captors by telling stories, of monsters and their maidens, high knights and low Jezebels, wise kings and evil sorcerers, who all lived and died in magic lands beyond the farthest seas. She told her stories in a sing-song voice, all the while accompanying herself with a small drum and large rolling eyes. Some say that even when she was delivered and stripped down to sobriety, she spoke in tongues and showed powers of divination; the other women said she is a witch—and kill her! The men laughed at this charge and argued that—look at her—she was too valuable to destroy. Maya did not fit the ethnic norm for local women—most of whom were short, thin, with black hair and eyes, and olive skin. Maya, in contrast, was tall and fair, almost ponderous in her fleshy but powerful physique, with long brown hair and pale eyes that burned whenever she would fix her gaze.

Dracul greeted her with his usual gesture of having a sharpened stake placed in the center of the room while indicating that she would soon be set upon it. Rather than cringe and plead as was the norm, Maya walked to the stake, ran her fingers across its point and said, "As you wish; but first I want a child with you." She then turned to Dracul full-face, and while removing her clothes murmurred, "Don't worry; I will make it happen." Maya performed this service much as young Gavagus had once done—with variety and persistence. But unlike him she did it with a show of pleasure, and with infinite patience throughout the time it took to get results. Her ministrations so pleased Dracul that he began summoning her every other afternoon, often keeping her through the dinner hour into the evening.

After a time, Maya told him that she was pregnant with their child. He believed her—to his own surprise—and ordered Vorkur to bring in Western doctors and make sure that everything necessary for a perfect birth would be provided for. They had a son and soon thereafter another one, but even when far-advanced in pregnancy, Maya perservered in her task of caring for Dracul. As he became more and more immersed in her relentless care, he softened for a while, ordered fewer of his subjects killed, lost interest in his torture-chamber, and talked to Maya about the ideals to which a king's power should be dedicated. This disturbed Vorkur (who knew everything) but he did nothing, knowing how dangerous it would be to meddle in a king's—that king's—infatuation. He also thought he was reading his master right: there was no nobility lurking in that soul, neither was there lasting love; Dracul would soon tire of the complexities and scant rewards of progressive government and return to his repressive self—but he would become more politically agile as a consequence of the weekly discourses on The Good and The True that he enjoyed with Maya.

It turned out that Dracul did dump Maya on one cold winter's day. He had been drinking secretly for some time, and when he saw that she had noticed he began to swell with rage at the thought that since that woman came, he, a king, could no longer do as he wished. He didn't kill her as Vorkur suggested but first had her whipped, and then delivered her into Count Luciano's hairy hands as a reward for faithful service.

It is written that when he did this, Dracul-Gavagus lost his chance, the last he would have to redeem his soul and generate a noble history for his land. But scholarship was rudimentary in that ancient distant land, so the story as it comes to us is a mix of shards and patches, fragments of a tale told for entertainment on a winter's night. It is also written that Gavagus was eventually killed, poisoned by Vorkur who could wait no longer for his chance to rule. Vorkur, in his turn, was torn to pieces by the nobles he thought would support him. Maya, in one sense, was lucky to have gotten out, for Luciano shielded her from the violence; he indeed had come to love her—she fit his hands so nicely—and he also depended on her judgment in planning for the battles that now raged between all and all. Of their children, those she had with Luciano as well as with Dracul, little is known—the monks and scribes having fled or died during the hostilities.

In the general terror following Dracul's death, the kingdom split into factions which warred for a hundred years until most forgot what they were fighting for and a shaky truce brought intermittent peace. The leaders of the various factions, languishing in a situation of checks and balances but also fearing change, became susceptible to the influence of the western church whose emmisaries preached that salvation—peace and prosperity—was only to be found in a reunited kingdom under God with a monarch who is a direct descendent of the first king, the noble Dracul. They had sent out monks and scribes to find such a one, and eventually a heir presumptive was identified—a Flemish citizen, a pleasant educated man now working as an an art dealer in Amsterdam. He is known to be perceptive, said the monks, for he represents the better artists of the period, particularly those concerned with a pantheistic intercourse—not strictly orthodox, but permissible in that liberal clime—between landscape and the figure. He also is successful as a businessman.

This settled the issue. For the rural and divided people that was Dracul's legacy, the possibilities of a united kingdom were so attractive, and the alternatives so dreary, that a determined effort to entice this notable young man to relocate and become their king, was mounted. Wladislaw, for that was his name—a westernization, the monks pointed out, of Dracul's 'Vladimir'—accepted the offer rather quickly, and did not ask many questions. The art business in Flanders was actually not too good, tastes having shifted to still-life painting of a highly mannered sort which Wlad found to be lacking in soul and wanting as regards the body.

The prospect of a coronation, after so many years of gloom, brought a singular conviviality to the discussions of the planning committee. Although representing different nations and conflicting interests, they all agreed that the new should be based upon the model of the old, for Dracul's coronation had gone down in folklore as the best that could be given, bringing happiness to all who stayed alive, and satisfying both spoken and unspeakable desires. When Wladislaw was consulted on this matter, he thought that while it is dangerous to disappoint all those who hungered for a replay of the Impaler's ascension, he would not accept the sadistic practices that had marked the pageants of that time. But Wlad was sophisticated in matters of deception, his father and his grandfather having both been artists and art dealers—so he knew more than anyone in that backward realm about illusion. He suggested an alternative to the committee—that they transform old reality into new art, in a way that would turn his actual coronation into a modern simulacrum of the first one, into a pageant which provided symbolic equivalents for the original episodes of pain and pleasure that had celebrated the coronation of old Vlad. This transformation of life into art, and art into life, would then become the theme of his entire reign.

Wladislaw was an entirely different king from Vladimir. Having been educated abroad, he had nuanced appetites and a taste for obliqueness in their satisfaction, and he felt it important for delicacy to become a factor in communal desires. He also was a pragmatist in other social matters; he denied that blood was the solution to all problems, just some, and he encouraged the army to concentrate on finding ways in which to feed the poor. The aura of intrigue and revenge, which had enervated the country for so long, began to fade. For the task of governing, King Wlad brought in faceless functionaries, academic experts from the west, and charged them with running things with as much fairness as could sustain efficiency. The nobles were given the duties of commemorating holidays and celebrating such notable events as coronations, and of putting on shows concerning the latest trends and fashions—for Wlad was aware of the political influence of noble well-placed women. To the long-term courtiers, he assigned the task of writing fifty-year plans for the future of the nation—for the king assured them that he appreciates the wisdom born of long experience. During his reign, Wlad took steps to teach his subjects how to watch rather than do, how to think so deeply about the act that the anxious muscles used in sword-play stopped contracting, and the blood would then flow to more ruminative places. Wlad had chosen a monastery-trained punctilious man named Vorkur—a descendent of Dracul's first prime minister as his majordomo—in charge of all attentions paid the king. In order to demonstrate his dictum of artistic substitution, King Wlad asked Vorkur to accept castration—under the best medical conditions—so all would know that the king's affairs were being managed rationally, without any residue of self-seeking.

This second kingdom of Dracul, if it had been left to the development of its modernized potential, and been surrounded by nations believing in free trade, would have lasted a thousand years. It did flourish for a while, primarily because the other nations could not distinguish the theatre pieces with which it celebrated its warlike past, from its present military power. But artistic styles, when they become familiar, are more easily separated from their subjects, and it was inevitable that some shrewd dictator of a hostile nation-state would spot a redundancy in the various pageants and would risk an all-out assault.

Wlad's kingdom was quickly conquered. His cavalry, all bedecked in bright red invincible uniforms, charged the opposing cannons—but the uniforms were soon stained by a darker, more prosaic red. King Wladislaw and faithful Vorkur managed to flee by making their way to the harbor and paying half of what Wlad could rescue from the royal treasury to secure two places—the queen having been detained in her chambers by the opposing avant-garde—on the last ship to sail. Eventually, they made their way back to Amsterdam, where Vorkur had a son, an artist, from an early marriage.

Walter had known the later sections of this story when he first began the pilgimage to find his origins. Indeed, they were his reason for going. Now that he was thinking of returning to America, he approached these last episodes from the other side, through the sequence that led up to them, and he found that reading from early to late was not the same as having read from the present to the past. Either way there were gaps in the record, and sections of both journeys took him through a chronological (and ontological) wilderness.

The way to the past he found more frightening, as there is no assurance that beyond a certain point there will be anything at all. So we must give him credit for undertaking the journey to retrieve an undocumented past. Having courageously passed the point beyond extant description, and having arrived at a place with no history—neither in space nor time—he was committed to live as a blip that had no record. To survive, as he knew from certain arcane studies, requires writing both him and place into space and time—thus enabling him to stay or leave. Being bold enough to first take residence, he could then construct his future life.

The way back has its own dangers. The threat of nonexistence is not one of them, for the documentation of both ends, past and present, is complete. But finding the intermediary roads between competing realities is not easy: Just because you're a contender for the completion of your line doesn't mean that past connections can't be missed, that someone won't bump you off the plane—so that you are stranded in a place which uses time but has no history.

When threatened by such predicaments, Walter had little recourse but to reach for his facility at invention. When he left his wife and his professorship to search for origins that would make them incidental, he knew nothing but the stories in the poems he had read in class and the lies he told his wife. His journey to escape a life of mere invention and sordid lies brought him, more than once, to a void encircled by the howls of knowing beasts. In defense, he thrust the heroes of his fancy into that fearsome place, and as heroes will, they overcame its perils—so he always found a path leading to a town where he could spend the night. Now, having gathered up these many pieces of his past (in discussions he called them "aspects of himself-as-other") he planned to return along the path leading to his newer, more realized, more honest self—not any other.

Of course, his flights of wishful thinking had to be restricted by the need to connect, without too much straining and in good time, with the next recorded episode up the line. By this time, Walter had grown some; he took a quiet pride in having succeeded so well. He had learned the mountain's fierceness, been a player in a play, survived assassination; and he had matched his fragile sexuality against the power in the hairy hands of Luciano—his new-found relative—and had won. Now he was travelling through the late-middle period of his heritage with the object of finding how it leads to the present—but this present, he was confident, would not be the same as the one he left.

Maia had listened to him tell the story of King Wlad with a distant look upon her face—the expression worn by people who believe that everything works out as destined. Walter felt his power increase with the telling of each episode, but he knew that in Maia's eyes this accruing status was also destined. Nothing that he himself had done deserved her praise, only the fact of his being who he is and doing what is pre-ordained. So Maia's love for him remained impersonal, for it was directed at Walter of the lineage of Dracul and not at Walter the meek and needy, the one who wants, in a warm and wet way, to be loved as his mother once had loved him. This obliqueness in their union was painful but not unfamiliar to Walter the long-time illusionist; and so he accepted Maia's attentions as she offered them because the art in them was good enough, that is, together they were closer to reality than he, by himself, had ever been.

Walter and Maia approached the issue of conception circumspectly. Oh, they began to make love again immediately after the last reading, but in ways that were preliminary—warm-ups and rehearsals—for the engagement of the destined egg and sperm. The count watched them from time to time, making small noises when they were particularly apt and leaving when they remained too long in place; but they ignored his presence, for his history, unlike theirs, was outside of destiny and so would end with him.

One day, Maia invited Walter to join her once again upon the mountains, in that self-same spot where her curiosity about her flogger-manque was first satisfied amidst the dusty rocks and the insects. But this time would be different. The rocks would still be there; but it was too cold for insects. It was the proper time to conceive a child. Walter prepared himself like an ancient warrier of his tribe. He sat naked staring at a wall and pushed his thoughts as low as they could go, past his lungs into his stomach and from there through his bowels in search for that detour that would deliver them to his gonads instead of following his excrement. He washed and wailed and wiggled his fingers to make dirty shadow monsters; he pulled upon his pecker, not to masturbate—no, he had to save himself—but to exercise its resilence so that it would be up to the task ahead.

Finally, the time came. Walter wrapped himself in a warm cloak, grasped his staff, stroked his balls, and with pilgrims' feet ascended the hill beyond the town. Maia was there when he arrived, sitting with her arms around her knees, looking much like the boulder that supported her back. They exchanged trivialities about the weather, and the cost of travel as if someone were listening in on their conversation. Then Maia reached beneath his cloak and grasping his hesitant member, began the rituals that would certify its mission. When he was safely inside her, Walter began the usual thrusting, but found the motion quite different from the in-and-out that had been all there was to it. This time, there was no going back; each thrust began where the other had left off—he was getting deeper and deeper into a place that had a beginning but seemingly no end. He began to ejaculate soon after entering, not like the popping-off of past orgasms, but in a slow and steady stream that also seemed to have no end. His pleasure was a steady glory more appropriate to the orbit of the heavenly spheres than to coitus on the dusty ground; Maia crooned songs of her native land and moved beneath him like the tides that follow the urgings of an autumn moon.

When the last light left the valley to touch the peaks above them, Maia pushed Walter from her—not forcefully but as firmly as was needed—and then resumed her replication of a self-sufficient boulder. Suddenly, with wild laughter and thundering hoofbeats, a group of horsemen came upon them, seized Walter and made preparations to impale him upon a freshly-sharpened stake, saying such ribald things as "He's kind of tight, grease up his bung-hole so it'll go in easier." As the point began to penetrate, another group of horsemen galloped up led by the young soldier in flowing sleeves brandishing his familiar rapier. They set upon the first group with swords and pikes, singing all the while the popular "Peoples Hymn of Liberation." When the battle had been won and the rescue completed, Mannie, the affable director of Theatre in the Castle, appeared in his beret followed by two cameramen. "Splendid, splendid," he intoned, "some excellent footage there, got it all from the beginning. My, you two fucked for a long time, but there's a big audience for that sort of thing. As for you terrorist guys, you could have taken the impalation a bit further, he was only beginning to show that he was scared; but never mind—another time. Maia darling, as ever you are the queen, whether in the flesh or on the silver screen—such talent, such grace; and Walter, you surprised me, yes you did—you weren't as clumsy as before—keep in touch. Now, bring out the wine and let's all celebrate a bit before we have to make that lousy trek back to the village."

Walter insisted that he and Maia leave before the others—his lower self being in such turmoil that his front and rear were totally confused, due in no small part to the mixture of oil and vaginal fluid that indiscriminately surrounded them. As they walked, he ranted: "If we're to stay together, we have to get out of this place. Such boors, such low-lives, complete lack of taste! And now we'll be seen humping all over Europe—perhaps New York as well—and what really hurts is the thought of my ass-hole winking out over all and sundry—some things are just too private to be made into art." Maia said little; she was quite prepared to go because she had it all inside her now, the latest documented version of Dracul, so it didn't really matter where she went.

The three—the count went with them—took the low road to the harbor and set sail for America. They had considerable luggage, for Maia wanted the costumes from her various roles, her party gowns and hand-shaped furs, slippers and snow boots, her manacles and the embroidered childrens' clothing that the old country is famous for. Walter packed a trunk full of the documents he had found in the castle archives; true, these belong to the state, but he felt he had a personal claim to them, for they were really, as he had interpreted them, about him. Anyhow, they won't be missed—no-one reads history any more. The count, with just enough brandy to sustain him, took his cape, his dirk, and a change of underwear.

Walter's spirits were high; he had survived the customs of that strange land and was returning, not as a supplicant, but with family, progeny, and source-material for his future work. Having no further sexual obligations, he was able in the course of the voyage to outline the book on the Draculian Saga. 'Vlad the Impaler' was a title that he was sure would interest a publisher. Of course, in addition to the material at hand, he would have to provide a socio-historical context, trace the interplay between geography and ideology, and strike the proper balance between vivid description—particularly of the gory episodes—and the distanced viewpoint of a relativized ethics. But he had, he was sure, the professional skills to do all this.

Before Walter could achieve his expected renown, there was an interim period of some unpleasantness—part-time teaching in a junior college, a squalling baby in a fourth floor walkup, the count drinking up most of what he made as a night watchman, and Maia's studied indifference to everything but the child, little Vlad, who from his earliest day, terrorized the toys put in his crib. But the book was finally published, and the reviews, while cautious—Walter had no major reputation—were sufficiently positive to draw good audiences to his talks at literary meetings, generate requests for articles from prestigeous journals, and finally, get him an invitation to teach as full professor with promise of an eventual chair at an elite university for wealthy high-achievers in central Massachusetts. Distinguished Walter (now "Walter Draculski") lovely monumental Maia, and feisty little Vlad made a fine addition to the campus community; and the count, now to all appearances Maia's slow brother, was given a job with campus security. All this brought to mind a statement attributed to another successful esthete, to wit, "If this isn't making it I don't know what making is," which Walter took as his own.

His rank and reputation secure, Walter's teaching began to disregard departmental course descriptions and become more personal. He was developing a theory that evil can be transformed by being turned into art. It was not a new theory; Nietzsche and his French admirers had produced a considerable literature on that topic, and the flitting in and out of culpability by singing and dancing while bad-assing was a staple of counter-culture. But Walter had a new twist, and increasingly he involved his students in its formulation. "What if," he would say, "during our Tuesday class, I smacked one of you in the face—you there, Johnny, wih the glasses—and shouted 'that's what you get for being stupid,' how would the rest of you react? And more—what if, after a few loud minutes of 'I can't believe it' and 'tell the Dean,' Johnny, glasses askew, stood up and he and I shook hands and laughed and said it was a test to see how you-all would react? Would you be disappointed that the incident wasn't real or would you be happy that it was only art? Aha! And here's another one: What if you were to see a film—a really tacky one, badly made, completely unconvincing—in which the heroine, so-called, goes through an ordeal of torment and rape before she is shot to death—and then you are told the film is a documentary, made by gangsters to celebrate their social deviance? How would you react upon a second viewing? Would the aesthetic flaws take on new meaning—a different value? Aha!"

Walter went on this way from semester to semester, presenting conundrums that illustrate the interpenetration of art and life, recounting some of his own adventures and past confusions on this issue; and for his graduate students, arranging showings of "Sex and Terror on the Mountain"—somewhat edited—which Eduard the director had kindly sent him before releasing it for European consumption. Walter made himself a bridge between intention and appearance, disbelief and believability; he argued that the psychic barrier separating truth and appearance was as porous as the formal one between reality and art. In his course description, he proposed to "overcome all such dichotomies, and thus permit the unfettered mind to entertain the multiplicities of identity." Walter soon became a cult figure in the university, and although some faculty objected, a burgeoning enrollment (Walter drew students from far and wide, mostly from Scandinavian and Third-World countries) ensured that he could teach exactly as he wished. Indeed, there were students who shook their heads and rolled their eyes, but most of those withdrew mid-way in the semester—"business majors," Walter would note with aplomb.

There was one student who came to Walter before the start of classes, earnestly told him of his faith in Christ and his dediction to righteousness, but said he needed to test that faith and dedication against other points of view—which he understood were plentiful in Walter's course. "Welcome," Walter said, "we have no fixed viewpoint but examine all beliefs in the light of our subject matter. Your contribution will no doubt be very valuable." Walter thought a virgin would soon be despoiled—but the young man, Aloisius Poniatowsky, the son of a local minister, stood and argued from the first day: "There is good and there is evil, right and wrong, a blessed path and a sinful road; one leads to God and Heaven and the other to the Devil's Hell. Sing or burn—everyone must choose." The perserverence of Aloisius, although it would be a good book title, irritated Walter nonetheless; he found himself somewhat embarrassed when he spun his tales and presented his quirky films. Actually, Walter was waiting for the boy to confess that all his remonstrations were first attempts at making art, and that he had now overcome his binary ways—but he didn't.

Other students in the class moved beyond their early irritation into outright anger; they had come as freshmen to submit their most foul desires and reprehensible appetites to the scrutiny of an approving ideology, so that they could graduate both educated and untouched—and here is Aloisius with his either-or. They told him that what he said was only his opinion, but it was evident that he believed his opinion to be firmer than theirs; and he wasn't moved in their direction by the logistic examples valorizing equivocation that Walter threw at him.

Then rage appeared: "How dare he thwart us," the students cried; "his protests are sapping the vital juices from the class." Poor Aloisius Poniatowsky.

The incident that made the university a byword in the annals of crime in higher education took place late one fall, near Thanksgiving, when the moon was full and the night was dark and silent. During the investigation, passerbys had recollections of cries and other noises, but no one thought to report them as they came from the direction of fraternity row where such disturbances are commonplace. The deed was not discovered until early morning, well beyond the curfew hour, by a co-ed who was trying to reach her dorm unnoticed. In a grassy space near the parking lot, she saw a thing on a stick, which appeared to be a large popsicle with extensions—no, a four legged spider at the limit of its web or a monkey sitting precariously on the end of a branch—maybe it was a new advertisement for a graduate thesis in modern dance.

Although well equipped with teen-age imagination, this co-ed mostly valued reality, so she overcame her fear of breaking curfew and started walking toward the apparition which grew larger as she approached. She stood before it in the light of the moon, and she saw between the dark stripes of shadow that it was an effigy of a human form, propped—no, impaled—upon a stake. She reached to touch it expecting to feel a plastic surface—papier mache painted to look like flesh. But there was a dark red liquid seeping down the wood of the stake and it smelled, a serious smell that anyone would recognize. She began screaming and didn't stop even after the campus police showed up. What they thought was an effigy, crudely made, of a man impaled upon a stake, was the anguished body of a boy whose impalation was in earnest. It was the body of Aloysius Poniatowsky.

The students responsible—perpetrators as they were later called—were easily found and readily confessed. On their lawyers' advice, they said they had been mesmerized, brainwashed, by Professor Walter's teachings, and they thought that the impalation would corroborate his thesis of the perfect fusion between art and life. It would be their final project illustrating the truth they had learned from him—"that everything is nothing that it is."

The incident made the front pages of the New York Times and Washington Post, and became an issue in the coming elections. Although the investigation and trial showed signs of being lengthy affairs, public reaction was immediate. Excoriations of permissive education, decadence in art, pernicious secularism, and intimated pederasty, shared the public stage with detailed stories of abuse in the childhoods of the students involved—Aloysius getting less space in the papers than his killers.

Walter was suspended without salary and went into seclusion; Maia came out of her lethargy to berate him for jeopardizing little Vlad's chances for success, and for his arrogance in taking on a role which was not destined. The count stopped drinking but said nothing. One day, Walter's body was found in a vacant lot near a downtown pizza-parlor which he had frequented, where he could take his solitary meals in peace. His neck had been cleanly and instantaneously snapped, which puzzled the police because they could not imagine hands strong enough to do the deed without minimal signs of a struggle. The murderer was never found. Some say the police were not enthusiastic in their search, and others say it was an opportune ending to a messy affair which was threatening to embarrass important politicians, ruin the university, give the town a bad name, and polarize a simmering dispute between left and right that both sides had tacitly agreed was best kept unresolved. In any case, the affair receded from public attention although it did eventually become the subject of a movie based on a book Maia co-authored with Mannie, the director.

Maia, the count, and little Vlad quietly left America and returned to the castle in the village where they had once been happy. Maia took up her theatrical interests again, and with the help of Walter's annuity, produced plays that were more timely, although perhaps not of higher quality, than those she had once acted in. Count Luciano now drank only in the afternoon spending the mornings teaching his adopted son—whose shoulders had grown surprisingly broad—all he knew about the treacherous arts. And so their little Vlad, a handsome mix of his mother's fairness and the darkness of an older lineage, was given all the advantages of an only child and became a force in Balkan politics.

XIX

Anna

THEY LIVED IN A large house on a small farm near a moderately sized village in Southern Poland. The father, Wladyslaw Belcarz, was a man of substance, a civil engineer, and so merited a home of scale and appointments with room enough to accommodate the three children of his first marriage and the one, perhaps more, to be expected of the woman who was soon to become a new mother to the first three.

Although he did no farming, Wladyslaw preferred to live in the country. The old trees and weathered boards of the outbuildings, the animal sounds that announced the beginning of each morning, the clear air and smell of hay—all these diluted the oppressive news of perilous and uprooted times that he had to face in town. The benign landscape of Poland's plains and forests provided Wladyslaw with an ideology as well as a profession. When he drew his plans for roads and bridges, projects that were optimistically being commissioned by his newly free country, Wladyslaw was more than ordinarily sensitive to the needs and conformations of the land, the age of the trees, and the balance between what is made and what is left alone. Beneath his once-a-week Catholicism lay the stronger belief that closeness to the earth and its animals are salutary for a tranquil soul and therefore indispensable to the proper development of young children.

His three daughters were plump and pampered; they spent all their free time in the fields and barns, vying with each other in the work they did while tending to the animals and taking joint credit for the progress of the crops. Wladyslaw's pleasure in watching their antics was tempered by his sense that it would all soon end—as it so often had in the past. He feared that his new republic would again suffer the degradations that the major powers periodically visit upon the Poles. He had childhood memories of horsemen coming in the night, of waking to the sudden noise of hoof-beats and of listening, from deep within his covers, to the voices that moved between curses and moans . They would be gone by dawn, and during breakfast he would listen to his father's explanation—which Wladislaw himself never quite believed—that these devils rarely harm the gentry and professional classes because, you know, we are a dependable source of money— but they do treat the poor peasants with an unseemly brutality.

In recent times, the gentry, with the help of friendly powers, had become quite effective in molding the peasants into a reasonable fighting force, and so Poland was able to shake off the tyranny of the old partition. The newly formed government presented itself as guardian of a noble heritage— the protector of rich and poor alike. This heritage, while richly celebrated in festivals with rediscovered patriotic music, was quite incidental, as even the patriots realized, to what would await them in the larger flows of history. To be sure, Poland is an ancient civilization; it is one of the earliest sources, in the Slavic West, of Christian belief. But its later history is mostly one of subjugation—and it was only recently, just a few short years ago, that the Great Powers had decided, for reasons of detente, to give Poland another go at independence. However welcome, in the ordinary political sense, this event only heightened the pervasive sense of doom among the better classes—to which Wladyslaw subscribed.

These elegant and lettered souls looked out, as they had in past times, at their flat, fertile, but defenseless plains, and their patriotically unformed peasants—not really good soldiers—and they were again overcome by the chronic mix of powerlessness and yearning that had always been their way. Still, they were not despondent, for they prided themselves on both their critical sense and their romantic flair—the former a trait borrowed from the French, the latter from the Russians. What they now must do, they all agreed, is to take on the crude habits of their sometime rulers, (for only a limited time, of course), so that the Poles could for once help win a war and make something of this unfamiliar gift of freedom.

The Belcarz farm was situated an hour's ride on horseback from the village, far enough to maintain the separation between town and country, but close enough to schools and shops to satisfy the family's daily needs and enable them to keep abreast of the latest news about the condition of the nation. The family did not themselves live off the land, although there was a vegetable garden tended by the children; and there were horses, dogs and cats, a favorite calf which with many tears was replaced every year, and a venerable rooster who spent the coldest winter evenings indoors, parading among the cupboards and upholstered chairs as he would do in the henhouse when the weather turned warm.

The rest of the estate, the actual farming and care of livestock, the repairs to the barns and fences, all this was left to the peasants who lived across the fields in small huts that belonged to the estate. Wladyslaw had read some Tolstoi but did not share the master's ideas about the nobility of peasants. He had little interest in the lives of his workers, for he found them rather dull brutes who regarded nature only as a stubborn adversary that matched their own reticence in providing more than the minimum needed to sustain their lives. Despite these sentiments, Wladyslaw was more generous than most of the other landowners. In exchange for labor, he provided his peasants with the huts in which they lived and some land to cultivate for their own use, and on Christmas Eve he gave them each a small bonus. But in his routine dealings with them he remained quite aloof—certainly, he made no attempt at a transcendental meeting of souls nor did he show an interest in their daily affairs, and he never inquired about their health or their children. In truth, the peasant classes seemed quite oblivious to the spirit that he saw pervading all the civilized world—discarding inert matter, dirt sweat and dullness—as it moved along.

During his university years, Wladyslaw had studied Hegel, and he followed the great man's doctrines on this point: The time of brutish confrontation with the earth for sustenance has receded into the past; and the present (everything that is entitled to move forward) will give to those aware of spirit's march the relief of contemplating the bounties of nature at a cultivated remove. So Wladislaw sought to maintain a proper, progressive, balance between the potential of his land and the other ambitions that he nurtured, and he encouraged only those enterprises whose image and flavor would enhance his social status. Of course, his plans changed in keeping with the play of politics—but this only added novel content to his garden parties and syncopated the older rhythms of planting and harvesting.

Wladyslaw was then in his middle years, good at what he did and well respected in the community. He had been blessed with a proper Polish physiognomy—bright blue eyes, a stalwart carriage, and a certain litheness, bordering on the dance, that he would display in social gatherings. His standing as an engineer had risen to where his approval was required for every public work—road, bridge, aqueduct—being built in southern Poland, a region which so far had escaped the turmoil further north, the fierce battles between white and red that had followed upon the independence granted Poland after the great war.

Although he appreciated his fortunate position, Wladyslaw was not a fervent patriot; he had little use for slogans or elections, for he saw political conflict as a predictable consequence of unavoidable economic inequalities. While he, himself, did not come from noble lineage, he had worked hard to become a trusted caretaker of state-owned facilities, and this achievement gave him privileged access to the people with the manners and delicacies he admired. Before the war, when Poland had been divided among three great powers, the region in which he grew up had been under the dominion of the Austro-Hungarian Empire—without a doubt a more benevolent and enlightened realm than the other occupying countries, Prussia and Russia. As a young man, Wladyslaw had learned an important lesson at his father's table—that good conversation, the best wine available, and discreet shows of allegiance, did much to ease the barriers between victor and vanquished—and once a degree of intimacy became established, further developments of mutual benefit would follow.

Wladyslaw often missed the clandestine pleasures of the old divided days, but he prospered in the new united Poland. As his duties expanded, he was required to travel to neighboring regions in order to negotiate such matters as contested water rights, inadequate surveys, and other matters of importance for a young republic. The Germans, who bordered Poland on the west, remained angry over the terms of the armistice and the loss of territories. But their actions were predictable. They would conduct their affairs according to rules inscribed in ledgers kept in the inner archives of the state and thus not subject to public scrutiny. But a lack of adherence to these rules, however well intentioned, was inevitably met with hostile reactions—not addressed in the rules. So Wladyslaw took great care in always presenting his negotiating position in terms that permitted a number of interpretations.

Despite these practical difficulties, there was something admirably picturesque about the Germans that Wladyslaw was drawn to—their excellent posture, shiny boots, waxed moustaches, and the gleaming spikes atop their helmets. They seemed to leaven all their practical affairs with a highly formalist aesthetic. There were the ever-present marching bands with everyone saluting back and forth throughout the day, then rounds of jovial toasts extending into the night and only sobered by the piercing bugle-calls of early morning—all the noisy glittering panoply of a beloved empire. This, toned down a bit, could surely be a model for the new Poland.

The Russians were another matter. In his dealings with them, Wladyslaw felt as if the most sullen and brutish of his peasants were on the verge of usurping his land and gaining power over his very life and soul. All of his commitment to universals—truth, justice, the good, and so forth—his expertise in planning public works and his refined tastes—for these barbarians, this mattered not at all. Ordinarily, Wladyslaw was an expansive man, friendly, voluble, interested in the ideas of others, quick to suggest new ways of getting along, and pleased to share drinks, compliments, and conversation into the night. But with the Russians—not the monarchists, of course, but these new red ruffians—he shrank into himself, said no more than was needed for his mission, and made sure to leave before their malice would focus on him, before his evident aversion to the spittle sprayed on him in conversation and the foul taste of the Bolshevik vodka, was noticed and would mark him as an enemy of the people, or worse, a spy.

Actually, Wladyslaw was a spy, an ineffectual one by modern standards—but then, the new Poland had little experience in such matters. He had been recruited because of the ease of travel his profession gave him, also because of his interest in the benefits that would eventually accrue to a successful spy. His mission was to find out how soon the Reds would invade: The Poles were sure, given the bloody fate of the Czar combined with the Polish preference for Western customs, that the Russians would soon sweep over them in a great cacaphonous wave. Despite these fears, the Poles continued to maintain the public stoicism that had served them well in the centuries of their various partitions; their present concern was that when the time came, the world would see them put up a good fight, see them again as dying nobly in battle—unless, of course, they, the Poles, were sufficiently forewarned to convince another major power, a more friendly one, to overrun them from the other side.

It was Wladyslaw's task to find this out. When will the Russians invade? So he peered and poked and assiduously reported the disorder and destruction to his contacts; but he saw no evidence of invasion—no massing of troops or supplies. Rather, he noted the incessant speeches urging on the revolution, the arrests and executions that followed soon thereafter, and he also noted that he never could find out who actually was in charge. They all dressed in dirty shapeless clothes, those ugly Bolsheviks, men and women alike, and there were no displays of rank or office. Yet, they all seemed to know where to go to work each day—even though nothing much was being done—and those arrested never asked by whose authority; they simply turned away and vanished. Wladyslaw could not get any agreement on the matters he had ostensibly come to negotiate—at what boundary the jurisdiction over a road would change hands, or how to approach the common problem of raw sewage dumped into the rivers. The Russians, he noted with some puzzlement, were not interested in such matters.

So Wladyslaw returned home and reported to his masters that either the Russians had all gone mad and would soon kill each other off, or that one day they would get an order from someone they did not know, and they would then invade Poland and kill us all. Of course, he noted, we must consider that there are enough Russians for them to kill each other for some time, and still have enough left for killing us. As for the Germans, Wladyslaw saw no immediate threat from them; they seemed too preoccupied with the refurbishing of internal ceremony, rather than to to covet Poland once again—although he had heard rumors, nothing consequential surely, of some turmoil in their cities.

Wladyslaw was relieved of his clandestine duties with thanks for his bravery and a polite acknowledgement that he had done as much as a good citizen could do under the circumstances. This tepid result would not, he knew, lead to the social triumph he had hoped for, but it permitted him to relinquish his duties gracefully and devote his full attention to family matters. While he was away, his first wife had taken ill and was now near death. This deeply saddened him of course, and he asked his children that they join him in maintaining dignity and composure in the difficult days ahead. Wladyslaw had never had a strong or singular experience of love, and the nearness of his wife's death did not give rise to feelings of grief—just to a deep discomfiture. He became more quiet than usual and took on a melancholy look, but he quickly resumed the task of overseeing the estate. In his orderly way, he apportioned his energies between his ailing wife, his children, and his land.

Each night before dinner he would spend time at his wife's bedside describing highlights of his previous journeys and recounting the recent gossip he heard in town. His wife had never gotten much else from him than a benign but distant affection—even when they made love. But then, she had never known more, except for one brief time, with any other person. So it did not seem too strange to her that her husband should not approach her dying in a manner other than their usual way. Indeed, the sameness of his manner comforted her, for it suggested that everything would somehow continue to be as it had been; and the regularity in the sequences of gossip and storytelling masked the inevitable progression of her disease until it reached its end.

Her three daughters—their ages at that time were five, eight, and twelve—sat beside their mother through the days, spending rather more time with her than their father thought appropriate. But they hid their fears from her as he had cautioned them, and they emulated his show of steadiness by recounting stories of school-events and describing the mischiefs of their pets. The youngest daughter, Marya, was dark-haired and had a round face; and her short strong body could usually be seen astride a cow or a horse, for she had found the common language that made them trust her. But she could not cope as well with humans, not even with her family, for when she rushed at them with noisy delight about some little thing that happened in the barn, she was told to behave herself and quiet down; then she would go back to the barn and sit silently with the animals—her father had voiced concern about her from time to time. But now, she sat with wide eyes by the bed of her dying mother reciting the names of her animals as in a litany, and she would lace each string of names with a word of praise or admonition for the animals she had named. In later years, she would leave her animals each morning to join the black-cowled women who remain in church after the mass is said, scattered in the pews, reciting their rosary beads.

The oldest daughter, Antosha, had been given a good deal of authority by her father; she was the surrogate son who transmitted his instructions to the peasants and acted in his behalf by keeping an eye on her sisters and ordering provisions for the household. This authority created a bond between Antosha and her father that the mother did not have. This daughter emulated the father in the deliberateness of her voice, her upright bearing, and her indifference to the play that occupied other children of her age. In school, she was the tallest in her class and was appointed monitor, a role in which she made use of the balance between kindness and severity that she had learned at home. She was quite good at arithmetic and sewing but had little interest in literature or history. Soon after her father remarried, she emigrated to America, to New York City, where she had four children—three daughters and a son, with a man who died soon after the last child was born. She seldom spoke of him, did not remarry, but worked at cleaning offices in the financial district during the depression, and eventually became a supervisor.

Anna was the middle daughter. She had her father's grace of movement and her mother's fine bones and fair good looks. She was precocious, bright, a flirt and teacher's pet at school, but was given to depressions marked by fits of rage for which no-one, not even she, could find any good reason. She had become her mother's confidant through the years, and the two spent hours reading from the many classics with which Wladyslaw had balanced his library of technical books. The literary recitations were often interrupted by confessions of intense feeling which were then diluted with bits of memory and stirred and blended in the murmuring of quiet intimacies. But despite the conflicts in this interplay—the unseemly passions mingling with the hallowed memories—the readings continued on, each one furthering the stories of the previous one; and through them Anna learned the finer uses of her native tongue and smatterings of other languages ancient and modern. She also learned how other people, both her mother and the literary ones, could feel in ways that were more powerful and less shameful when they were taken out of the body and set in words. As she grew, Anna began to see that she was different from her sisters and her schoolmates. When at first she talked to them about the important things she had learned from all the stories, they ridiculed her. Anna's mother was concerned that her favorite child might unwittingly expose what the mother had hidden from the others all her life. And then Anna would be harmed, the victim of a revenge the mother knew would be taken were the others to find out how her daughter also is inside. So Anna's mother tried to teach her how to live two lives, a public and a private one—of managed affairs and of consuming feelings—and how to benefit from each by using the other as a shield.

But the mother was now dying and the lessons were not yet complete. Anna asked her mother not to die, to wait awhile, until such time when it would not be so hard for either, a time when neither one would care as much as now. Her mother asked her not to say such things in front of the others, and then she said she knew the feelings are not yet finished, but that she had not managed her affairs too well. It was the last lesson she could give.

Wladyslaw waited the proper amount of time before he took another wife. There were many women to choose from, for he was handsome, still vigorous and quite successful—a widower with property and an established family. The woman he chose to marry was only eight years older than his daughter Antosha. She was dark, full-figured, clear about her likes and dislikes, and given more to ordering about the children and the servants than to introspection. Her forthright sexuality helped Wladyslaw put aside his habits of wearing night-clothes and sleeping under his own coverlet, and he soon began to arrange his professional and familial responsibilities so that they would all be disposed of shortly after supper, about the time he and his new bride would retire for the night and resume the play of hands that sleep had interrupted the night before.

Wladyslaw's daughters, having lost their mother, were now losing those aspects of their father that he had divided among them as they grew. They were afraid this losing would get worse, and after bedtime Anna would recount to them the many stories her mother had read to her about stepmothers, and how wicked they can be. Antosha was the first to leave Poland, not so much because of Anna's stories—for she had little use for the truths of fiction—but because the new wife had taken from her all her customary say over the practical part of family affairs. She asked her father what he wanted her to do, but he answered with the evasive hostility he would use when dismissing an unneeded worker. So Antosha left, the basis for loving her father having been removed, and she spoke very little of him from then on. When, years later, she received notification of his death, she sent a condolence card bought at the corner drug store.

Before Antosha left for America, she made a pact with her two sisters that they would follow her as soon as she could learn enough about the new country to ensure that her sisters would find work, a place to live, and a chance to meet Polish men whom they could marry. Of course, Wladyslaw was not too happy with the prospect of losing first one then all his daughters—although he was soon to have a new child, hopefully a boy, and more would surely come. So perhaps the daughters' leaving, after all, was for the best. In truth, Marya had become quite opaque to him, a demi-creature, more and more like her animals, saying nothing on her own and responding vaguely to his questions when she could not move fast enough to avoid his presence—best to send her where her well-being would be someone else's concern. But he felt differently about the prospect of Anna's leaving. Anna reminded him—she had from her earliest years—of the wife he had now lost. She walked in the same constrained yet supplicating way, and exhibited the same slight tremor, the reddening, when in his presence. But she also brought back memories that now embarrassed him—which he no longer wanted.

His wife had said to him in those early years when they were first married, that she would give herself over to his wishes, and then she would press him to tell her what he wanted. He enjoyed this sense of power and began to look for desires he had not known he had—but however variously they tried, his satisfactions lacked what she, despite herself, could not bring herself to give him. So, early in the marriage—he remembered that it was really at her instigation—they developed a ritual in which Wladyslaw would beat his wife, once a week, with the finely oiled expensive belt he reserved for his best trousers. He always made sure to stop the strokes when her buttocks achieved the redness that just precedes a breaking skin, and when her moans escaped her determination to control them.

There was of course no malfeasance on her part, no neglect of duties that provoked this; it was rather an attempt, yes, freely undertaken as they both agreed, to span the distance between them, to see if the gap would close as she came to fully feel the hurt—a hope that the noise and smell of it would join them together as completely as their vows demanded. She did enjoy it—but only when she took it as a secret penance for her failure to feel what the dark-tales described. Despite the weekly whippings and the years of child-bearing, she remained as distant from Wladyslaw as on that first day when he brandished his hairy organ for her astonishment and delight.

In her last days, she tested the distance from her death through the inexorable approach of the pain now deep inside her, and found she could maintain it as she had the monotonous strokings of the whip. And so she kept these pains as she had the early ones—within herself—and only showed them to Wladyslaw when she became angry at him, finally, just before she died.

Wladyslaw did not beat his daughter Anna; he never even struck her in the way that fathers do—although he liberally cuffed and slapped the other daughters. Anna's sexuality had been gotten from her mother and she used it to gain a little girl's power for controlling her father as she saw her mother do; and she succeeded, for her father had no defense against his recognition of the mother in the daughter. After his new marriage, he increasingly kept Anna at a distance, but even with a new young wife and the open lusty ways that followed, he was loathe to send her off together with her sisters. He feared that she would take with her the memories of her father's centering his passions in another body—before the first one had found peace within the memories that Anna had—memories that no one else could know.

Anna did leave Poland, together with Marya, to join the older sister in America. Wladyslaw did not much protest; his new wife gave him reason not to. The night before the sisters were to leave, he put on his finest cloak and led the family in an exuberant discussion about the opportunities—status, wealth, children, yes, his American grandchildren—that were just waiting for those of us who now take the chance that God is offering. And, surely, as God knows, we will someday soon, all of us, be back together again.

The American decade of the 1920s, the one that preceded the Great Depression, is remembered as "the roaring twenties." Many factors contributed to the noise: The machines of industry grew large and multiplied; motor cars and their honking horns followed the new buildings that were invading Thoreau's sylvan glades; the public babble was enhanced by loud sound of jazz and the murmur of exotic languages from across the seas. But most of all there was the promise of success, made manifest through competing dins of argument and celebration, and the equally noisy concerns with money and fame.

Some folks, particularly in the mid-west, considered these concerns excessive and the images reprehensible, a consequence of the unchecked immigration of unwashed, ungodly, foreign polyglots. And, in truth, some of these concerns could be traced, in hindsight, to our European allies in the Great War who, as a token of their gratitude for America's intervention, sent us certain of their corrosive ideologies and a large sampling of proponents. These were offered in good faith, as a simple salve meant to lubricate the dryness of skin and psyche that the Europeans noticed was inflicting the members of the American spirit—as evinced by the boorishness in its Expeditionary Force.

However, this influx of aberrant ideas and needy people was quite transformed by the richness of our domestic practices. The American versions of money-fame-love were then more democratically apportioned than in Europe (where they are traditionally relegated to the well-born or depraved). But here, fame and love could be enjoyed by anyone with money, influence, and the social graces to hire eager rubes and foreigners—anyone willing to work at scratching another's itch.

Certainly there was a winnowing process, even in this land of liberty and equality—for not everyone could reach the level of participation required—many Americans stayed home, dry and low, and deep within the sticks and stacks. But for the other restless souls, the quest for these exotic but increasingly fashionable joys became so fevered that it took the form of a pilgrimage. So it was quite appropriate that these post-war pilgrims, the deprived natives of rural America and the unwanted castoffs from an exhausted Europe, would all head for the new center of desire, New York City, a magical island ringed by a confluence of rivers and connected by glowing bridges to the outer lands.

When they arrived, with their armpits stinking and their bags filled with dirty clothes and optimism, these pilgrims, red-faced rubes, sallow skulkers, and bearers of unpronounceable names and sometimes noble origins, began to molt the coarse original skin that had helped them cross the outer lands and water—a shedding that would reveal a soft new skin hidden beneath, the more responsive skin that would help them survive in the changeable climate of this center of their new-found hope. They had stoically endured weeks of monotony, the steady stench and awful food, but despite all they had heard, they were not quite prepared, when they finally did arrive, for the brazen cacaphonies of their new world—the brightest lights they had ever seen and the loudest sounds, whirling around them without stop, jiggling their bones, and scrambling their thoughts. They had been told back home by those who had reasons not to make the journey, that all the blinking and booming is just the American way—a constant froth of celebration on the surface of waters that are underneath opaque. They want you to take off all your clothes and swim with them.

Some were indeed put off, young girls with scarves tied beneath their chins and old men in heavy shoes, the ones who did not want to come but made the crossing because they could no longer stay at home. The young ones could not just open up their arms and legs, wade right in, slide and shimmy to the dances, and take their chances. The old ones wanted comfortable shoes and a place to think about dying. Both the young and old, all those frightened ones, had been told that someone will meet them when they arrive. So they did not talk to strangers, they turned their backs upon the din and brightness, and walked with the practiced shuffle that back home had kept them safe. Some were trundled across the rivers to more dim and muffled places, outer boroughs where relatives who had come before and were the sponsors of their journey had made a place for them to stay. Others came to the empty places ringed by the predators who live off new-found freedom.

The Greenies (those who had just arrived) scurried through the ominously geometric landscape (avenues in names or letters and streets in numbers). They wore layered combinations of their native dress—the bulky skirts and heavy coats carefully brought across in steamer trunks—which they mixed with the poorly made but up-to-date American clothing—sold in shops on every corner, that they were spending their carefully hoarded exchange dollars on—because one had to look American to become American. (Citizenship was on everyone's mind). The greenest Greenies could be identified by their ignorance of the symbolism of clothing combinations. In time, the men would learn that to wear a fur hat and a frock coat offers a different view of things than does a fedora and a vested suit; that the difference between brown and black makes for a needed separation between day and night as well as life and death. The women had more difficult codes to master. They had to decide between the expense and status of a fur-trimmed coat and a simple but servicable one of cloth, between a shawl tied beneath the chin or close-cropped colored hair, between the least and most of ankle to be exposed in different places, and whether all shoes should be suitable for snow, or look as if they were too delicate to enclose a peasant foot—the sexuality of shoes was new to them.

Antosha, Anna's older sister, lived in Brooklyn, in a neighborhood of Slavic immigrants—Poles and Russians, Christians and Jews—who knew each other through the history of ancient enmities, but who were comforted in this strange place by the familiarity of their old hatreds, and were joined together by the perils of their adopted country. The streets they came to live in had no grass, few trees, and no sign of what they had fled—the oppressive relics of an ancient, green, and noble past.

Antosha had waited patiently for the sisters during the long hours of the passage through customs. Then, without any comment beyond the customary Polish greeting and a perfunctory kiss, she took them with their heavy trunks directly into the subway where, swallowed by the roaring beast, they sat through each clattering stop and start—Anna trying to shield her body from the rub and poke of strangers, and Marya seeing the devil leering in every vacant sleepy face. After a time, the train came up into the light, a moment of epiphany—and across the miraculous bridge across the moonlit waters, truly a symbol of God's concern—but then down again, as evil will spread across the world—down into the darkness. The rattling tedium and Antosha's continuing calm slowly overcame the sisters' fears and they leaned against each other, contracting into a sleepy lumpy bundle of sweat-stained clothes. Suddenly Antosha's voice broke through their dreams, saying they must wake up now, the next stop is where we get off.

It was a five block walk to the building where Antosha lived, a long way for carrying trunks. Some passing men turned to look, but there was no kindly peasant with a horse-drawn cart to carry their belongings. Antosha did not rush them but neither did she show much sympathy—for this is how it is over here, she said later—everyone carries what they have, and complaining only leads you into trouble. The apartment was on the third floor, a railroad flat as it was called, a kitchen with a window on one end, then a procession of small rooms leading toward a larger window on the other end; the single toilet was behind the kitchen sink. Antosha had been using the largest room, the one the landlord called the living room, as a bedroom for herself and her young son, and her three daughters shared the small room next to the kitchen. But now, with Anna and Marya having come to stay—at least until they could find work or husbands—the three sisters (Antosha's daughters) would have to sleep together in their mother's room, and the other sisters (Antosha's sisters) would be moved to the daughters' room. The son would sleep on a cot in the hall. There was no delimiting period indicated for the limit of Anna's and Marya's stay. They were welcome—and there was no sly suggestion otherwise—for as long as it took them to get on with their lives. But that is what, starting tomorrow, they had to do.

After they had settled in, Antosha began taking her sisters to the Tuesday evening gatherings of neighborhood Poles held in the basement of the local Catholic church—other nights being reserved for the Irish and Italians. The concept of the melting pot held for the Sunday masses but did not extend to social gatherings. Though joined together in their Catholic faith, each ethnic group would wage sporadic war against the others, using as their weapons, memorials of native saints and historical triumphs, superiority in cuisine, and from time to time, fists and broken bottles. The priests in their weekly sermons said nothing about these social gaps between the faithful, thinking no doubt that common beliefs had never stopped a war back across the seas, and that, in any event, the journey of the soul to paradise is an individual matter.

So when the beer was served and the polkas played on Tuesday night, the sisters met only other Poles. Sometimes a vagrant Czech or Lithuanian appeared, smelling for some new food—but not any of the others, none of the dark-skinned or fair but freckled ones. The men who came to talk and look and drink and dance were thick and heavy, with calloused hands and smelly breath—much like the peasants who worked on the estate in Poland, the coarse ones that Wladyslaw had told his daughters to avoid. Here as there, the men worked as laborers, there in the fields, here indoors among machines. The smell of their bodies was more pungent here, more sour, for there were no winds to refresh the air and cool the groins and armpits. So before the men appeared at Polish Night, they would hurry from work to their sinks to wash away the residue of the day, and then change into a clean shirt and their Sunday suit. Work did not end until after sunset, so the men always arrived at the hall late, after the women had settled in, after the band—accordion, violin, and saxophone—had begun to play.

When the men got there, they first gathered near the walls in the darker corners and talked about the kind of women they preferred. Each vied with the other in performing extravagant gestures and hearty laughs, for these were telltale signs for the watching women across the room of their prospects as a husband—signs of self-confidence and an affable nature. After a time, the men would wander in twos and threes to the womens' tables to present themselves with formal greetings and small stories that they would tell in the most flowery Polish they could muster. This was an obligatory prelude to asking for a dance, first with an older woman, then, if that went well, with one of the younger charges.

In the public space of Polish night, the older women looked upon the young unmarried men as if they were visitors from a distant village—at once untrustworthy and comic. These women, aunts and cousins of the newcomers, looked and talked among themselves, each remark and its rejoinder accompanied by a knowing laughter and directed through oblique glances at each man in turn. This was not a pleasant laughter; it was accusatory—suggesting secrets discovered and weaknesses laid bare. It was meant to diminish and belittle the men, and to unsex the reasons why they had come. Such accusations, the women knew, are the base upon which relationships are built; they ensure the separation between the weights of soul and body that enabled their own mothers to resist giving pleasure to themselves or their husbands—a resistance they well knew, and the young ones should practice—for it provides women with strength and maintains the stability of marriage.

During Polish night, the older women controlled the rites of passage: They measured the promise of each supplicant against the merits of their own husbands; they tried to thwart loose infatuations and they moved to promote the sensible unions that would most resemble theirs. The younger women were given little encouragement to display the ornaments they had surreptitiously saved to wear for such occasions—for this, silly girl, is not the point! You have been shipped here, from old to new—but it is all the same. The river separates us from an evil that is faster and lighter here than the devil we once knew. Our salvation is the same, however— it is to marry and make families through which we shall be protected, old and young, from the nameless seductions that beckon all around in this ungodly land.

The most suitable men for pairing were identified by the older women through a whispered summary: You see over there, the tall one in the brown suit leaning against the bar—that is Stashek. He's a little drunk tonight but he has a steady job as a machinist, goes to church and lives with his mother—a good boy, quiet and hardworking. The short one next to him is Walter; he's finishing high school at night—he's ugly but they say he's very smart. Avoid Edvard—the silent one in the corner—because he is ugly and not smart -and he likes to preach about a god we do not know.

It took but three weeks, three meetings, thirty-some polkas, for Marya to settle with a man. It was somewhat unexpected, for her sisters had for so long noticed her distant and unfocussed eyes—and she did much finger her rosary—that they feared she would end up a nun. But she met her Franusz by walking to him—without any prompting or hesitation—across the dance-floor, to where he was sitting silently among the other men. He was silent because he could not talk as fast as they, and because he knew they would not listen to what he had to say. He was as short and squat as Marya, made shorter by the curve of childhood rickets in his legs. His face became more earnest with each drink, and when Marya showed him that she would listen, he talked about the oppression of the working classes and the need for a revolution, or at least, a need for stronger unions. He told Marya that he preferred drinking to religion because whisky did not tell him lies, although it made it harder for him to tell the truth. But, listen, after the working classes take control there will be no need for whiskey or religion. Devout Marya looked at him as she once did at unruly horses, admiring their spirit and indifferent to the content of their protest. They both saw they were in the space that separated them from the others, but they also saw they could inhabit their little space together. Greatly relieved, they married within the month. Antosha looked at all this with bemusement, for she never could understand her youngest sister, never get behind the blank openness that covered she knew not what. But Franusz was a hard-working man; everyone had only good to say of him, except that he read the Polish-language newspaper "Glos Ludowi—The People's Voice"—frowned upon by the church for its radical politics. He also would remain serious when it was time for joking.

Anna did not meet her husband at Polish Night; she had expected to find a way of living in America that would recognize the same distinctions between the fine and coarse of things that had made her village so orderly and peaceful—the kinds of distinctions, after all, her father made when he chose her from among the other daughters as his special child. She knew she must not look for her father in a husband—there were holy reasons, a priest once told her, to keep such things separate. She did need someone enough like Wladyslaw to show her that the difference between fine and coarse would not be too great—would not eat at her for the remainder of her life. But here at Polish Night, the men were just like the peasants she had known back home, noisy smelly stupid clods. Here—in America of all places—they were free to breathe and sweat on her, and tell her how fortunate she would be to wash their clothes, cook and clean, and have their ugly babies—just like the servant women in her father's house. But she was here, not there—although she still did not understand just why. She also knew she could not find the other way and marry a man so cultured that he would abandon her as her father did, when her unsteadiness made her a hindrance to his life. Each week she came more and more to dread these nights; it was reasonable to go—but she could not stand the sweat and whiskey-breath, and the bleary leering at her body outlined by the thin cotton dress that Antosha made her wear.

One Tuesday, as they were all dressing in preparation, Anna began to scream, accusing Antosha of treating her like a cow for sale, of raising her own children to be as coarse and stupid as the others, of not having told the other sisters what it would be like over here. Antosha heard her out; Anna's tantrums were not new to her. In Poland, years ago, Anna had used them to gain the father's private admiration and install herself as his favorite—so she could sit with him at his desk and tell tales about how the other sisters avoided their chores. But Anna's tantrums would not work here, no matter how much she screamed; here she had no choice—marry, get a job, or out on the street you go.

Nevertheless, Antosha was worried; Anna paid little attention to threats or pleadings, and her outbreaks had become more frequent and more erratic; every evening it now seemed, through supper until sleep, they had to listen to her complaints and accusations. She did not seem a sister any longer, but a half-tamed animal that had somehow gotten in to their small apartment—and she was seriously upsetting the children. So Antosha went to ask the advice of a relative, a second cousin who was a physician, born in America, the most successful member of the family.

Doctor Michael wore vested suits, expensive ones which permitted him to keep his jacket open and expose the tailored vest draped across his considerable belly. He listened to Antosha's recitation of Anna's waywardness with some dismay, for he had noted Anna's frailty when she was brought to him for the complete examination that he performed for new members of the family. He had examined Anna delicately, marking her blushes and the touch of sweat that developed on her bare skin when he probed her and demonstrated how she should care for herself. He did not then notice that there was a perverse edge to her responses—her stretched out beauty had obscured the darker hints of her instability. But now as Antosha described her ravings, he did remember that her conversation, especially when on family matters, had swung rapidly between aggression and love, between fast laughter and the beginning of tears. He also remembered thinking, even then, that what Anna needs is an older professional man, one who would be both stable and firmly supportive—a Victorian type, he thought, a man who would encourage Anna to remain a pampered and obedient child even while she bore their children.

With these recent episodes, however, it became clear that mere advice, even coming from the trusted doctor, is not enough. Over the years, he had seen many members of the family, charging them very little, and showing a sincere concern for their wellbeing. But with Anna, Michael doctor cousin connoisseur, felt an obligation that transcended the medical and familial. He could not let this girl shatter into screaming fragments; she was too beautiful and too vulnerable—he clearly remembered his first examination of her—indeed, she would be a considerable prize for anyone who could understand the larger sources of her needs, someone such as he, but unencumbered by a family, a pot-belly, and a comfortable practice.

So Doctor Michael suggested to Antosha that he make a dinner party, a familial joining to which he would invite some friends, male familiars who, just as he, had surmounted the Polish stereotypes and were established as professionals. This was indeed a gracious gesture and it was greeted by the show of gratitude it deserved. Michael received the effusions modestly; he knew the steepness of the path between his terrain and theirs. He also knew that he would enjoy the drama of the evening, for he had a well-developed voyeuristic sense that was not limited to the interplay of body and disease, but extended into events of other colorations where the participants knew less about the forces motivating them than did he.

The doctor picked his dinner-guests carefully; they had to be suitable as providers for Anna's welfare, and they also—this was more difficult to judge—had to provide a contrast to her fair plumpness such as would make their coupling picturesque in different ways. So he cast her suitors by constructing a tableau of opposites which would complement her: the bony and the rounded, a prominent nose against the yielding underbelly, clean fingernails rubbing up and down the softly freckled skin. Michael exercised much care in his selection; indeed, his first duty was to Anna's well-being, and to ensure it he went to a privileged place, a dispassionate perch, from which he could guide the dynamics of the choosing and which, later, would reveal how it is going with the one she chose.

Of course, she would be frightened after the first coupling; she would feel shamed and exposed, and this would bring her back to tell him what she could tell no other person—because she trusted him, her important doctor kindly cousin. She would obey the advice he gave her about the things she had to do and let be done to, that would transform her awkwardness and revulsion into the unexpected sensations that he told her were her due, that befit her grace and her many attributes. Doctor Michael would not charge her for her weekly visits, but he would insist upon a detailed and scrupulous accounting of the happenings day by day.

One of the guests that Michael invited to the dinner was an undertaker; he was a sturdy man with a large moustache whose droop he used to convey sympathy and condolences without having to re-contort his face for the particular grief of each succeeding client. Like the doctor, he was partial to vested suits and the gold watch-chain which attests to the prosperity of those whose profession it is to tend the dying and the dead. But the undertaker, his name was Tomasz, did have some faults; he walked heavily, with splayed feet and an asymmetric slouch, and his hands while soft and large were quite hairy. Also, he talked about people he had known without specifying whether he met them when living or enbalmed.

Michael also invited two members of his social-athletic club, first generation Poles with developed talents for business. One coordinated routes and shipping schedules in a trucking firm, the other was a specialist in medical supplies whom Michael counted on to provide the better products at discount prices. The first, who liked to be called Freddy, had the expanding physique of a one-time athlete and a cheeriness of face and voice that was pleasant, informal, and undoubtedly contributed to his business success. At work he knew everyone's first name, was adept at decreasing friction between the shippers and the drivers and, as he confided to Anna and the other guests, would soon be slated for promotion. In social situations however, especially in lengthy dinner parties, Freddy fared less well. His amiability would stay intact no matter where the conversation went, and the other guests would eventually feel smothered by its lack of content and would begin to look for ways—trips to the bathroom, even rudeness—to escape from its monotony.

The name of the other young executive was Gerald, but he preferred being called by his middle name, Peter. His parents, in the interests of assimilation, had named him Gerald, but there was no Catholic saint so-named. Peter was fixed in his devotion to Catholicism, a devotion that began during his first teen-age troubles and grew steadily ever since. He once thought of entering the priesthood, but he was not particularly attuned to other peoples' cares and problems; and a priest he once confided in told him he could best serve God by finding a vocation that suited his excellent memory and fondness for detail.

Doctor Michael had coupled Freddy and Peter to further his strategy of contrasts. Freddy was big, bluff, and loud, while Peter was small-boned, somewhat short, sallow, and indirect. During conversations, Freddy would gaze with loving candor into each and every face; Peter seldom looked directly at anyone when he uttered his carefully considered statements of medium length. Freddy was partial to sport-coats, brightly patterned shirts and loafers with tassles, while Peter wore single-breasted suits, white shirts with initialed cuff-links, and mutely striped ties. But both were concerned with the help that appropriate attire can give to professional success—Freddy energetically displayed the symptoms of his happiness, so as to convince others that he was always happy. Peter never seemed to be enjoying himself, or indeed, to find enjoyment particularly desirable—but he presented himself as someone who could be trusted. Freddy exuberantly ate and drank too much, although he was learning to discriminate between cheap and better wines. Peter did not drink beyond a courteous sip of the offered wine, and ate only a single serving of food, delicately refusing seconds.

Anna did not care at all for Tomasz, for in his after-dinner discourse on the formalities of death, she detected a whiff of bad breath. Freddy interested her even less; she thought him quite dull and knew that he would become fatter and more winded as he aged. She was somewhat intrigued by Peter, for she had never seen a Pole, not even her father, so clean and trim. But she could not imagine touching his skin, for it looked like parchment that would not bleed even if scratched.

The last of the guests that Michael invited to his gathering was a violin teacher who had no studio of his own but walked throughout the neighborhood, often covering considerable distances, to meet with his students in their homes. His name was Stefan. He was older than the others, the poorest of the three—an artist, of course—and he seemed an unlikely choice for Anna given his age, and the threadbare patina that the careful pressing of his clothes could not hide. But Michael, as a matter of training, was in this as in other matters, quite empirical. In his medical practice, he did not prejudge diagnoses, but typically included even the most unlikely candidates for consideration as causes in an illness; or, in this case, as suitable prospects for a young woman's fancy. Stefan the violinist, was not cowed by his rivals. Why, he had played for officers of the Imperial Court—princesses and politicians—not as a soloist, of course, but as the third chair in the second violin section. During his early years of touring in Russia, he did not adopt the rough Slavic histrionics of the other players, but took as his model a conductor, a Pole educated in France, who projected a continental fastidious that contained a touch of self-doubt. Stefan learned so well that he often became a target, especially during intermissions, of the coy remarks that ladies in the audiences would use to annoy their husbands.

Many years ago in deep Siberia, toward the end of the first world war, in a town called Harbin, the Tzarist officers drank the last of the available vodka, and lit their cigars with now-worthless paper rubles. Stefan, my father, was there, playing second fiddle in the orchestra of the caberet where civilization as the Russian nobles knew it was coming to an end. My father told me about the laughter, the jokes and the seeming indifference to their fate that these moustachioed, finely booted, sharp-sabered, terminally drunk relics of the passing empire showed as the cannon-sound of the Germans, and the early infiltrations of the Soviets, came ever closer. But this was a problem for the morning; the nights were reserved for the grand gesture—the challenge to the Gods who might otherwise think their bravery was wanting.

My father also told me about the battleship Potemkin. The incident took place in another town, Odessa, just down the coast—his troupe travelled anywhere they had a chance at getting paid. They had arrived there just when the mutineers on the ship, their Czarist officers overthrown, and their demands for supplies having been refused, decided to shell the town. "Boom" my father intoned, getting to his feet (he was already past eighty)) as he described his search for a refuge. Then, still on his feet, with his arm pointing to a different place: " Boom-boom, and I run to get under cellar steps." He then slowly sat down, and his story drifted off to descriptions of broken buildings and civic chaos. But he soon picked up the thread again: After the battles with the Germans were lost, the Tzarist army retreated and the cabaret was closed. So there was nothing for it except to get back to civilization—Moscow preferably—as the Siberian winter was setting in. There was a train they found that would take them—musicians and politicians, military wives and camp-followers, small business families and civil servants—in a southwesterly direction. No promises as to destination, but it was better than staying.

After some days, the train stopped at a small station, in the fields with no town in sight. This train, the conductor said, has been commandeered by the Bolsheviks—who will be here shortly in their own train; they are scheduled to link up and go on to do important business. Our conductor will be going with them, and he had no advice for the passengers except that they would have to live in the station-house until the next train comes. Many trains came by; they all carried red flags but they did not stop. The passengers fought, first gently then fiercely, for the best spots on the station floor.

My father, Stefan, was a mild man, but he also had to protect his violin—the only thing of value that he carried, which he hid from his uncertain companions. He had wrapped it, case and all, in some heavy canvas, secured by twine. He found a space for himself not far from the door, colder than the more interior spaces—but one that gave quick access to the trench they had dug for their toilet. But water and food soon became a problem. The water, with effort, could be gathered from melting ice—the food was a matter of what each party had brought with them. Stefan had some dried meat—jerky—but he had to give it to the communal pool, although they— those ruffians—did allow him to take a last bite. Nevertheless, they would all have starved to death; but a train did eventually come—which they were allowed to board after being subjected to pointed questioning by the new Bolsheviks. My father, for his part, raised the artist's flag, showed them his violin, and so hid his conservative beliefs. He had to play for them every night, which he really did not mind, although the cold and the lurchings of the train upset his fingering, which the others, those primitives, did not seem to notice

After that, the story becomes vague again. How Stefan eventually came to America I do not know. He once said that to pay for the trip, he had to sell his prized violin for a lesser one—the one I have played on for these many years—but he thought that no one would notice the difference in the new world. My father had little patience for stories. His language skills—in Polish, Russian, and English—were limited, and he also did not want to tell the stories of his failures (for that is how he thought of his life) to anyone—not even to his only son who did not, until later, after he was dead, ask him pointed questions.

Making the remaining world into a party during the winter of 1917 had a good measure of insouciance—the Tzarist officers joking over champagne as their world was coming to an end. It was not merely the failed campaign—the broken glass and bloody bandages around cut feet—but the realization that all that they had cherished would not come back—the pigs would occupy the palace, and music and poetry would be limited to patriotic songs and slogans. But if the officers were discouraged or afraid, they did not show it. What is required, they said, for the celebration of one's own demise is only that the last gestures show sufficient evidence (in the best possible light) of the glories that went before. After all, it's what did happen—not what will happen—that matters to the celebrants and to proper history. Stefan had great respect for the nobility; he admired the finery of their appointments and the bravura with which they met the increasingly hostile world. But my father was not like that. He emulated them as best he could, but he was after all, a peasant-fiddler. There was nothing of the past he remembered that meant much to him—he was interested only in what will happen next and how he can survive.

Stefan had been raised by women, his father having disappeared soon after he was born. He always was more comfortable with women than with men; men talked of prospects and profits but women knew what it was like to express feelings. The mothers of his present students were usually at home during the day, during the hours when he came to teach—the time the men were working—and the manner of the conversations the women had with Stefan was more important than musical progress in ensuring that the lessons would continue. Stefan the violin teacher was by now a familiar figure in their Brooklyn neighborhood. Most days, usually between ten and four, he could be seen walking the streets that joined his various students. He did not mind walking; he had walked much longer distances in Europe, between villages where they were scheduled to play when the trains stopped running, away from villages when the fighting came too close, and sometimes just from a place to another place when things and people became too difficult.

Stefan was tall and slender with good posture, and he carried the violin case in his left hand, leaving the right hand free to tip his hat to those who greeted him. He had appeared in the community some years ago, rather suddenly, and had distributed leaflets announcing his strong European credentials and availability as a teacher of the violin. The stories he would tell of his earlier life were vivid with historical places and events but vague on personal details. Some of the men thought he was a fake—he was too polite; and truth be told, he didn't play too well. No matter, the women said, Stefan is a cultured man, a handsome man, a musician who after all has played in the major cities of Poland and Russia, and who, no doubt, will soon join an orchestra here, perhaps in Chicago or Detroit—he often speaks of his many friends there—and he will leave us and our children the poorer for his absence.

No-one thought to question the propriety of Stefan's daily visits to the various households. Violin lessons were after all a symbol of prosperity, of an arrived and cultured family. Not every father could afford to buy his son a violin—and also pay for lessons. Not all children in the neighborhood could—as could his children—play Polish songs at the family gatherings to the applause of envious relatives. So it was unthinkable to read something else into Stefan's daytime visits when the fathers were away at work, unthinkable that the wives might be getting lessons too, instruction in the ways of the wider world, sweetened by reminiscences of intrigue, assignations, and revelry in the days when the great war was ending badly and no-one cared. The fathers of Stefan's students could not know how much more exciting was that world than the one they provided—even the one where there is meat on the table every day, and the weekly cleaning is done by the more recent immigrants. For us, they said, there are summer trips to the beach at Coney Island and celebrations in our living room with the family at Christmas and Easter, where we can see the children grow and the old folks die within the circle of their loved ones. What more is there? God has been good to provide all this in so short a time.

Threadbare Stefan, the violin teacher—a mere artist after all—never had so much as he has now, and surely, his visits provide entertainment for the children and also satisfy the mothers. For it is unthinkable for the uninitiated that by merely correcting the fingering of a scale, Stefan could reveal how it feels to be alone; and then through a wisp of remembered melody how it is to come together with another while both still struggle, for ancient reasons, to remain apart. Who could think, that by the tapping of his toe in three-quarter time, Stefan could demonstrate how the rhythms of coitus are strongest when the prospects of their future are limited to only a few more times.

Those men, those Polish husbands, who had come to South Brooklyn of America and had survived and prospered—for them to even think such perverse and foreign things—meek Stefan as vile seducer—would constitute a sin whose mere thinking, like all sins of thought, could then turn into a deed—a murder, perhaps, which was not impossible. After all, the Cossacks killed for much less. But if it were done, and witness borne, no one could deny the right of destroyer action against the secret sin of once-obedient flesh—and the duplicity of a seeming friend. But that would result in two murders! Although common in the old country, such an act, especially in American Brooklyn, would soil them all, reveal the origins—for all to see—of the stains on their souls and on their sheets. It would subject them to jail and, worse, to snickering in the bars! So best to give it no further thought, and look again at the women who appear, from time to time, in certain neighborhood establishments.

The wives—the mothers of Stefan's pupils—had a different sense of sin. The thoughts their husbands would not think were not as frightening to them as was the prospect of living in their unchanging lives. So these forbidden thoughts of change found their way into the afternoon sessions of tea and wine that they held before their husbands reappeared, in which the wives would grade each of their good fortunes against their discontents. In the later hours of afternoon, the wine helped them speak of imported desires. They would then take on a vocabulary of allusions which they had learned from magazines and movies, and which they practiced upon each other while their husbands were at work. They particularly enjoyed the freedom which speaking indirectly gave them, a speaking which became their secret language. Often, they would discuss the scenes in movies which showed foreshortened practices of the French and Italian upper classes during the vile old days, practices which they mostly all agreed no one could now engage in, certainly not in Polish Brooklyn, even had they wanted to, which of course they did not none of them want to even think about—much less do unless they had to.

Their husbands never gave any signs of understanding the delicacy of such things. The dirty stories the men would tell among themselves, stories remembered mostly from their youth, were coarse and to a certain point—the brutal joys of copulation—but in the retelling, the pictures of furtive groping had been lost—obscured through the years of sterile re-telling. Stefan had heard many such stories, scattered across the continents, but he did not tell them—not to anyone, not even to his son—only high-born people could do so without blame or shame.

At the dinner party that Doctor Michael gave to bring his second cousin Anna into a relationship with a good man, Stefan showed up late— "My apologies but my last lesson was on Garfield Place, and I had to first walk home and put away my violin and wash, you know." He said all this in a plain but accented Polish, and although the others were somewhat shocked at the unnecessary recitation of his toiletry routine, Anna found she liked his public joining of the clean and private; she could see the wash-cloth moving down his body as the violin rested on the bed. Stefan made some perfunctory comments to the doctor, gave his little bow to the other men—the small swift bending of the back he had learned in Russian cabarets—then he moved quickly to the empty chair by Anna's side and spoke to her as if they were continuing a conversation started earlier, elsewhere, perhaps in Poland or during some recent afternoon when they met in Prospect Park.

Stefan's life had not been going too well of late; the mothers of his students demanded more and more of the conversation with which he began and ended lessons—for which he received no extra pay. But then they often cancelled lessons on little notice, citing only the dull obligations they were sure would matter little to such a one as he, an artist after all, who did not have the burdens their husbands had to bear. Stefan heard the old familiar contempt beneath the tinkle of their fashioned Polish phrases, and he realized that once again his violin and his stories had been worn as thin as his shoes through the years of walking-talking. Each time this happened he had simply moved, as far away as possible, to where no one knew him and where he could begin his presentation once again, perhaps with better luck. But this time, he did not want to pack his large and heavy trunk and carry it, exchanging hands with his violin case, to the train station. He was now past fifty, he did not practice much any more, and the orchestras he had tried to join when he first came played music that he could not read nor understand—anyhow, there are so many young good fiddlers these days.

But perhaps he could put his age and memories of old Russia to a better use; instead of going off, he could start again right where he is, in Polish Brooklyn; and with a new young wife he would not be as dependent on the whims of those other women. His position as a husband—and even a father, he is surely not too old—would identify him as a person of substance, stability, worthy of respect. He could then move out of the furnished room and establish himself and his family in a large apartment with one room reserved for his studio where he would begin to practice again—it is surely not too late. Like a true professional, he would hang out a sign: "Stefan Krukowski, Violin Teacher." He would no longer walk from house to house; his students would come to him for their lessons—payment in advance.

Stefan had seen Anna a number of times, and had made some inquiries about her. He liked her large eyes, and the intimation that her sister found her difficult. She, like he, needed a new circumstance, and she would certainly ornament his presence at the various functions in the neighborhood, gaining him the respect, perhaps with a bit of envy, that would follow upon his attainment of a young, well-made, and obedient wife. He was quite prepared to love her in the appropriate way, for he badly needed someone he could have directly, whenever he was in need, without the incessant talk, the hand placed on the heart and the repeated gestures of devotion. But in his courtship of Anna he knew he must appear circumspect and dignified, projecting certainty about the future and the prospects of a comfortable and tranquil home—and yes, it is surely time, after all his accomplishments, that he now think of children.

Stefan and Anna were married in 1928, and Lucian was born in 1929, one of America's first replacements for the financiers who were destroyed when the great crash took from them all they had become accustomed to. Giant shock waves raced from Wall Street to extinguish the national ambitions of the land, and smaller players moved across the bridges to Brooklyn where they sent tremors into the ethnic neighborhoods—all of which severely dampened the lights that had once been shining new and bright. Stefan lost his students—not immediately, but through a steady attrition. His students—these hopeful scratchers—were able to study the violin because their fathers' employers had been convinced by experts that a small increase in wages produces a large increase in work. The economy was expanding in the 1920's and so the experts proved to be quite right—and their strategy worked particularly well when applied to the industrious, greedy but deferential, first-generation immigrants. And had things continued in this way, had not the bankers and politicians poisoned the wells of individual production, why, Stefan could have opened a small conservatory, hired teachers, and become the benevolent maestro whose "good, good—but more minor scales," would send the students into paroxysms of weekly practice.

As it was, even the most envied fathers in the neighborhood were losing their jobs, and the upscale values of the family that reached for music lessons and summer vacations at the beach was being replaced by low sullen drinking in the bars. After a while, Stefan had no more students, and there was no money. At first, Stefan could not accept that his gesture of commitment—which he had held off making for so many years—would so quickly come to nothing more than recriminations which started in the early afternoon, counterpointing Lucian's yowls, until he had to leave the house.

Oh, he tried to follow his last dream; he walked the streets as he had done before, calling on the families of his erstwhile students, telling the mothers how talented their children are, what a shame it would be to stop the lessons now, that he would be happy to reduce the fee, yes, how hard the times have become but how well they, the mothers, look. The children, especially now, should continue their studies, for no family in times like these can be indifferent to learning music which, as we all agree, is the quickest way to fame and fortune for a talented Polish child. And who knows, it might be Sigismund, or little Ralphy, who is so blessed—the next Wieniawski—such a pity, worse, a shame before almighty God, if one did not at least try, through a proper course of lessons, to see if there is a divine spark in the sweet child. I'll play a tune for him, a simple Krakowiak—look how he listens! Yes, he has a musical nature!

But soon he was no longer offered the obligatory cup of tea. No one—not even the mothers who were once most talkative and coy—would any longer want to listen; and he found that on his walks he was increasingly joined by peddlers of small trinkets, beggars who had not eaten for some days, and angry men who would take anything from anyone they could bully. It was clear: The violin has ended; it is done as an instrument for living, loving, posturing, pretending, surviving. Put it away, Stefan, pack it back into its case together with the other hopes you had unwisely let go loose when you came to the promised land, and think back to what you once knew well—that the awful is the usual.

Anna knew something was wrong. Stefan had not put the customary money on the dining table the last few friday afternoons. She said nothing; perhaps it was an oversight—but three weeks now? She borrowed money from her sister Maria, to whom she could tell such things without shame, and so there was still chicken on the table during the weekends, and at other times potato pancakes with apple sauce that she would make fresh with just a touch of cinnamon, and borscht with sour cream and pumpernickel bread. There was less meat than there had been the month before, but Stefan did not comment on the money or the food, nor did he say anything about the rent that would be due next week. More worrisome was that Stefan said nothing more about the plans they had made before they were married—about the large apartment they would rent, on a lower floor, perhaps facing on the avenue—and the house they would buy as soon as the good news of his conservatory would spread—the word of his pure Polish approach to studying the violin, an approach that in all of Brooklyn he alone could offer.

It didn't happen that way, as did not happen the dreams of millions who had thought that coming to America would be the final triumph of their struggle against the beast, a victory against the horsemen of their European fate which—can you believe it—they would win simply by placing an ocean between the two, the old and the new. When they first set foot upon this shore, the statue in the harbor told them—I swear, they insisted—I heard her say it—that they will all be given a new-world guarantee that never again will they lose all they have, the way their fathers and their fathers' fathers did, each time the collectors—those emissaries of the beast—came for the taxes and the tithes. Yes, these riders, who came each time just after sunset, were indeed offspring of the beast—for hardship cannot simply be blamed upon the dithering fools in government. No. Hardship comes from the evil one, no doubt—and yet each Sunday those other fools, the priests, counsel penitence and patience!

When Anna walked across the planks that led from the ship to American soil, she was sure that all the bitterness that had been given her to take upon her journey was not needed here—it could all be thrown into the dirty water—or better, deposited in the wondrous American trash bins that line the dock. Or perhaps, she thought that she would, if she could, return the bitterness to those who need it most, the timid ones who prefer to stay in Poland picking at their ancient scabs, cursing what they know too well, and praising in their prattle what they have learned nothing of.

The night before we left, they said to us—in their ancient knowing ways—that staying here it is better: The old misery is spread out evenly through the village; all share the pain our country gives us, but this—you must believe— is also God's way. In the new world, they warned, you would have to face your time alone—there is no sacred place, no ancient village, where you and yours can live and die together. It is said, that in America priests are mocked by Negro children, and that the Anti-Christ and Jews have been welcomed into the government. No, we will avoid the coming end and stay here as we should—and you should too.

Anna knew they are wrong—she prided herself on her courage. She could have told them that in the new world, even in Brooklyn, there is no Anti-Christ, only the chance to find new ways to what we need and do not yet have. Remember, back in Poland, the nobles lived high on the hill behind great walls and showed us nothing of their lives. They rode their horses past without a glance—not caring whether we live or die, or why we had been born. It is different here in American Brooklyn. We also have wealthy folks, wealthier here than there, but they smile and wave as they ride among us in their motorcars. Those who run the government ask us for our blessing, and also for our vote. They tell us that together we can change things for the better and they will help us when they are elected. Anyone who works hard and says these things can be elected. You must be born here to be president, but otherwise, there is no limit to what one can become in America.

Anna worked hard to understand the new ways, and she would lecture about them to her sisters until, confused and irritated, they would ask her how Stefan the violin professor was doing. They knew she could not tell them. Each week Anna hoped that Stefan would put a bit more money on the table—just for food and rent for now—but there was less and less. At first this only confused her—it was not how he had described their life. Then, looking at her sisters' eyes—the old country's eyes—she became frightened. Somehow, everything had stopped just when it was starting. Just the other day, it seems, there had been all those plans and dreams to talk about and fill a supper full of laughter—and the laughter continued on into the bed, where what Anna had to do would not be too unpleasant because of the happy talk of better times that had just preceded it. A few months after the marriage, she became pregnant.

Anna had learned English quickly, so she was able to read the stories in the daily newspapers about the great stock-market crash and the leaps from high windows through which wealthy Americans—like the ones who smiled and waved and asked everyone to support them—demonstrated that they, too, had pain. But all this seemed remote to her, a matter America would surely take care of—so she tried to remain cheerful, telling Stefan at the dinner table that this morning, in the Daily News, she read that better times are coming soon. Stefan did not answer her, for he had no faith in good news; but relying on himself had also come to nothing. He had made a mistake—that was clear. The mistake was to think that his charm and old-world manners would bring him success in this community of louts—these displaced peasants with fat wives and dumb children. And on this expectation—imagine, at your age, Stefan—he had staked a marriage with an ignorant young girl, and now—Stefan,- you will have a child.

But no, it was not a mistake and it was not his fault. This had happened to him before, when the powers he could not have anticipated swooped down—Mongol hordes from behind the mountains, Cossacks dispatched by the Czar, then the dirty Bolshevicki—and each time, they trampled on the hopes which he had lulled himself into believing. He responded now as he had before—he hid himself. But this time he did not look for caves or cellars—he had a deeper refuge. Stefan retreated back into the old age he had concealed when he first came courting Anna. Within this refuge, he was no longer a husband or an expectant father; he could become what he so far had wanted not to be. He became a grim and grey old man to whom little could be said, not even by a pregnant wife, without it being met by angry silence, or by an abrupt departure.

When Anna told her husband she was pregnant, he said nothing, and left without looking at her. Stefan was not a drinker—one way he could distinguish himself from the vulgar Poles, those who had never seen the Czarist nobles dance. And when he returned late that night, Anna did not smell the liquor she was actually hoping for, but only a stale nervous sweat—an uncommon smell for him—for Stefan took pride in cleanliness. Anna had hoped that he would for once be drunk so that she could comfort him and get him to tell her about the devils that were tormenting him. She had often wanted to do this for her father; she would have been so much better at it than that cold stolid woman he married late. Now she feared that she too had married a cold one. Surely, in troubled times people move closer, if only to join together in the lamentations, and so share the failure and its sorrows. But Stefan moved away as if he did not know her—or perhaps, she thought, he saw her and the child within her as the enemy, the cause of his misfortune. How could she not, at the beginning, at Doctor Michael's dinner, have seen through his sweet words and graceful gestures and have run from them as a bird from a snake. It would have been better for her to settle for the strong and sweaty arms of one of the boys she so disliked who tried to fondle her at the weekly dance. But now she was pregnant and Stefan had stopped talking to her, and there was no Friday money on the kitchen table.

So she went the other way—the only one she had; she went back to her sisters. The birthing of a child is too important to be put aside by a husband's whims or other transient things. Conceiving is indeed transitory—but, once done, it becomes a thing eternally present, understood by women and some others who know that cycles and renewals are unlike simple changes. Anna spent her time with her two sisters, walking to each of their houses on alternate days, and returning with what they gave her for the supper—not to feed that fiddler's gullet, they said, but for her to send the food through her blood into the child, so that it will be born God-willing strong and healthy. The food was important, but more so was the talking, the stories they would tell her of how it is to hold the child, and how quickly the boy—somehow they knew—recognizes you as his mother. The rest of it, they said, depends upon the father, and there it seems you will suffer more than you had thought—for you were always the best in school, the favorite of our father, the one the boys would always look at when you turned and twisted your behind during the dances at the parish hall.

Antosha raised the specter of a failed marriage before Anna was willing to admit it to herself. Her sister told her that it is not a matter of divorce, or an abortion God-forbid, but of finding a way—now that the fiddler has shown himself to be a wanderer, a man who cannot make a place for his family—to bring money to the table by yourself. Then it does not matter whether he stays or leaves—better that he leaves. Of course, you will have to go to work, as I have done now for twenty years—my husband died too young and yours has already lived beyond his death. But all that does not matter now. When you find a job you can leave the child with Marya during the days; it will teach her what to do when she has her own.

Anna did not talk about the unfolding of her pregnancy to others, not to her husband, not even to her sisters. She receded into the sensations of the stirrings as they changed throughout the sequence of numbered days, while she met her familial duties with only a cursory diligence—there was still food on the dinner plate, and clean clothing in the basket, but no longer with any pride in the accomplishment. She could feel the slow swelling with her hands and through the tightening of her clothes, and she now thought only about the life inside and her need for a dress that fits. Her larger sisters gave her the clothing that would shelter her, and they did so with a graciousness that contrasted with their gibes during the first months of her marriage—which were only the usual Polish belittling of fanciful expectations.

What Anna learned from this is that reality is only a coping with what comes next—and kindness, if it does come, helps. But by the time of her third trimester, Anna no longer knew what to expect from anyone. Expectations of pain, which had so dominated the early part of her life, had now become irrelevant—she needed to pay attention to her present. She was pleased by her first morning nausea, and she dwelt on the continuing moments of her pregnancy with a clearer happiness than she had had at any time before.

XX

Pain and Pleasure

Past and Present

There are many reasons why the past is considered to be better than the present. These reasons are often antiquarian and nostalgic—as in calls for the restoration of one's youth, with its early foibles cushioned now by greater wisdom and a bit more money—or in attempts to restore the monarchy, with the hope of a position, however modest, within the entourage of the new royal family—or for the prospect of living in a primeval Eden, replete with noble savages, a pacific lion, concupiscent nymphs and satyrs, fruit-bearing trees mirrored in a clear pond, and perpetual peace—or, if all the above fail, a better understanding of how the present will be after God has died and art has ended.

But there are reasons to desire a return to the past that are not nostalgic. These are in the relationship of past to present as regards the phenomena of pain and pleasure. We prefer pleasure to be timeless—ensconced in the eternal present—unless, that is, we confuse pleasure with the satisfaction of obligation—in which case it becomes a moral-political issue—perhaps a dilution of what we really want but dare not have. There is, as such, no "pure pleasure," but we can purify pleasure as it happens, although we cannot sustain it long enough to ponder it—to see if it is really pure—or not. We should not try. Pleasure is distracted by analysis; pleasure is to have. When it is over (are there good reasons for ending pleasure?) even then, it's traces can be seen—the sweaty sheets, the dirty dishes, the aromatics of sex and cooking mingling in the night-time air—good footage to document our "having pleasure."

Pain is different. We prefer to look at pain than have it. There is plenty of pain to see—in all of antiquity right through last night's shooting just down the block. Our vignettes of pain are most often revealed (to the pain-free) through embellishments—poetic, journalistic, historical—which offer images that suffuse suffering with the virtues of style, righteousness, inevitability: See the martyr's gesture that, in its extremis, greets the tranquil and forgiving light of heaven; look at the flag above the battlements that justifies the rage to kill. Watch the twitchings of the deer you almost missed, the day before the season started.

Pleasure documents in other ways—by mostly private memories. Some are reinforced by repetition, others stand out because they are affronted by disclaimers like: "What am I doing this for?" The success that pleasure has in public presentation (now a major industry) also makes it (periodically) open to the charge of immorality—of abetting, say, pornography, avarice, sensuality, and other such kinky pastimes. Hedonism, as we know, is a poor attitude for fighting wars- it is distracting. Yet, the pervasive enjoyment the enemy shows works well for them—it provides a focus. For us, pleasure is suspect as a way of undermining aggressive government—pain works better. Our images of pleasure—to be tolerated at all—must be coated with a redemptive distance—given moral purpose— before they can be shown to those who do not yet know the difference between dissolute and remedial pleasure.

The manifestations of Venus in Western art are all—as is Venus—"other." The images, of course, are fictional—idealizations of the fecundity of the Pagan Gods. The actual models, in the main, are but suggestive hunks of flesh that largely through male genius are transformed into the ultimately desireable and completely unattainable. Given these safeguards, Venus can be admired and desired and exhibited in all her splendor—think Rubens—for our reflective pleasure. With all our powers of mechanical reproduction, we have nothing that can compete with such images; we also have no overweening, yet God-sanctioned, lust. We have no, e.g., besotted elders who, at life's end, only wish to see Susanna naked—and can. But we do have lots of images for ongoing electronic presentation: Pretty skinny girls are hired to sell cars and jewels and bikinies—to all the multitudes for the unreflective pleasure of concrete impossibility. Pleasure, in any dilution—ideal, historical or coercive—is largely welcomed by those who like the fiction better than the act. But even this is not universally welcomed—especially by those who abhor the feelings of pleasure that their fantasies give them.

Contrasts

Pain is not relevant to our better natures, our deeper lusts, or our misfortunes—although it is acutely relevant to our feelings. Pain will prey upon the well-lived life, but it regularly visits the deprived and underfed. Pain is the great equalizer; it is witness, second only to death, to the end of life—however the life was lived. Pain must be borne where we are—there is no question of not having, or of tuning in later. When we are in pain; we cannot think of pleasure. Pain and death intertwine when we are past affirmation or denial, when we can no longer look to others or save ourselves. Pain is the harbinger of mortality's fleeting time although it takes place in the eternal now. Pleasure occurs there as well—but it does not have the same "now" as does pain. Pleasure must be imported—supported by the extensions of memory and anticipation. Pleasure is anti-death, in that, like life, it has indefinite times and modalities from which to choose. But the choice—and fulfillment—of pleasure is never clear. It is subject to interrogation even when in action: Why am I doing this? Who is this person? Will I get into trouble? Pleasure, even when intense, is more casual than pain. It is pro-life in the ordinary sense of life—which looks for pleasure while avoiding danger.

Pain is not like ordinary life; it is always special. It becomes present in a particular body, which relays it to a unique mind. This process is not consecutive but instantaneous—as fast as neurons deliver sensations. The realization of "being in pain" by the mind, is temporally the same as the fact of "having pain" in the body. Mind and brain, whatever their structural and ideological difficulties, have no option but to "share" their pain. The best that can be said of mind-brain and pleasure, however, is that the "sharing" is as variable as their dual constitution. Perhaps a chart could be made that shows the increasing difficulties of holding to mind-brain synonymy in the light of various experiences of pleasure—starting with the stronger individual excitations, then moving to such social pleasures as compassion, attraction, love, then to the "pro-forma" pleasures of digestion and elimination.

I do remember pleasures as fully, although not as clearly, as the many pains of my past—like the aching tooth just now, the clumsy fall last week, the enema back when I was a bloated three. We wish that our "right now" pains could be as past as the one when we were three. These old pains are bad enough in memory, but not so bad in fact as is my present toothache. Strangely though; the memories of my young pleasures are usually better than the ones I have right now. My early cavortings remain with me as pure pleasure; they have few of the snaky pains that invade most everything I enjoy these days—and the enjoyments I look for now, the ones that taste the best, also attract the most poisonous snakes—each with a different apple in its mouth.

Although some say that pain brings the righteous together through its revelation of their common destiny, I, for one, do not want to feel their (or even your) pain. No-one wants to be placed in direct competition with the Olympiad of contenders who have suffered most throughout the history of sensibility—the witches burned at the stake, heretics broken on the wheel, blasphemers disemboweled, deserters hung from the yard-arm.

I know my pain most acutely before it stops—any comparison with others' pains doesn't help. But pleasure can be more easily shared. Indeed, it needs testimony to give it presence; it is customary to signal others that pleasure in life is good. But when learning of another's pain, we—even the more empathic of us—will revert to remembering our own ascendancy over old pain (the fortitude that made us what we are today). Then we are free to offer the comforting words that we expect will give relief to sufferers: "Hang in there, it will be better soon."

Early Pain

We really don't remember our early hurts. A bully's kick to young and tender shins comes back more as humiliation than as hurt. I am too old now to be bullied, so a kick to my present shins would more likely come from a small dogmatic colleague with whom I have long disagreed. But this would evoke a different response than merely a return kick—perhaps breaking a bespoke chair over his balding head.

But the pain in distant memory is not like present pain; it shows itself less as anger at what they did to me than as guilt for what I did not do. It does not recall direct sensation, only its circumstances—when trying to cope with bullies down the block, or where I broke my leg, or the time my first bare date stepped on my toes.

But, still, we cannot but have memories of past pain—for they mark old occasions more strongly than do other of our memories. I remember the long-denied pain of a tooth-ache, long ago, which finally sent me to the dentist—where I faced the antique drilling machine with cables whirring over wheels and the small deadly drill-bit spinning, pinning me with its squeal to the chair that I was placed in. I knew that once I sat down, I was helpless. High and low and round and round it goes—wherever they need to situate me in their protocol of formal and beneficent sadism. But I was sure, even with the drilling, that some other side—the pleasure side—might yet appear. Yes indeed—when we are young it usually does: The nice fat nurse presses her big bosom onto my skinny arm, and helps guide the palliating needle into my gum—just a pin-prick—but the growing numbness makes moot the promise that I, in silent but righteous self-defense, had made—namely, to distinguish between what is occuring in my mouth and my need to mouth the nurse's all-enveloping breasts. The bald and owly dentist knows the difference but doesn't care—his thing is drilling.

I have other early memories of pain, of cuts and bruises while playing young games—small wounds that didn't much hurt until swabbed with iodine by disapproving elders. "Why do you play with such rough boys?" my mother cried. I, you understand, am Polish, while they, those boys—although we go to the same church—are Italian and Irish. There is a difference! They were large, and had yet larger siblings and heavy fathers. I had only a small mother. The alternative, as my mother advised, despite her Polish-peasant roots, was that I play instead with the Jewish boys, who were less rough and talked more, and whose sisters often had large breasts and sometimes came to walk with me when play was over.

Diseases whose hurting happens slowly are harder to remember than sudden hurts—they do not have the drama, but they linger across times. These pains bring periods of restless sleeping that occlude the time of living. They become the persistent dreams of hurting—the ones I dream when I am ill. They have me walking over high paths which lead to villages whose inhabitants do not like me—and I must go on and on until there is a place with no inhabitants. I walk through large buildings with many corridors that shield me from the the village noises. I have no idea where the buildings are, why they were made, and what is being done in them—but they are cleaner than the buildings I know—larger, built with stone and glass and steel. But as there are no people in them; their purpose, as I imagine, is to be empty. My dreams when I am well do not have such buildings; these healthy dreams always change, although they are collaged with moisty matter. Such dreams affirm wet pleasure; they are affirmative of life—which, after all, is mostly water.

My early illnesses began with a slight malaise before bedtime, and turned to fever and cough by morning. Then the doctor (my second cousin) would pay a house-call, and while snacking on my aunt's chopped liver, he would say "flu" or "pneumonia," and plunk down pills I was to take until they were finished—at which point we could all believe (and it did so happen) that I was well again. The sequence between ill and well became more uncertain as I grew older. I still had the occasional raspy cough, but then came sudden rashes and scary sores, depressions without apparent cause, bowel-and-bladder consequences of bad food and too-much drink—all of them by-products of my new maturity. I tried to hide my torments behind day-dream regressions to the dream-paths of an earlier innocence; but pain and blame were more intertwined in this later play of grown-up roles than when my early suffering had the comforts of chicken soup, a soft hand upon my brow, and Saturday confession.

Early Pleasures

Yet, despite all that the invidious world would tell me about what is to be done (so as to avoid pain, omit guilt, and increase pleasure) my temptations in the early times remained solitary. I came to think that pain and temptation are much the same, and that they both are of my own devising. This led me to sweaty nights and dry-day confusions about what I want to do and why I should not do it. But despite the ache of sin and knowing that the all-seeing eyes of the Virgin are always upon me, I continued to retreat to the solitary act because it gave less pain and more completeness—indeed, I now can say, more pleasure—than the park-bench fumbling at which I was quite inept.

Later, much later, when I had my own place and began to earn some money, I could distribute my collage of pains and pleasures into different corners of my kingdom. After all, I had a bed, a bathroom, a stove and running water—all of which I could pay for—a palace where I was lord, protected by a snarling dog and attended by friends and foes, loves and occasional thieves. It was a place where I could do large things with breasts that, at the time, were smaller than my nurse's, and I could work on my erections, which were nowhere near the advertised standard. But as we all, at that time, were being educated by Hegel to go beyond ourselves, we saw these lacks as mere delayed fulfillment—with great expectations for a synthesis of mind and body by the time we graduate.

But pain still lurked in my palace—it did not know its place. It had been muffled under layers of learned expectation about the sequence of an artist's life, but it remained unimpressed and did not disappear. Pain is ever-jealous of the new directions pleasure takes. Novelty first led me to the smelly bed of unwashed sweating and cheap wine. Eventually—yes I brag—it led to more perfumed beds and good Chablis—gestures of appreciation from older others for my newly learned enthusiasms. Yet, I remained afraid—despite how good this new world smelled—that, despite the few yellow bricks in front, I was really following the cinder road out back which leads, not to the shining city on the hill, but to that other place where, between shrieks and supplications, no one can escape the reversal of their fantasies—the place where pleasure is never (ever!) fulfilled.

I did try for some scholarly comparisons: Is there sex in Heaven and Hell? How does being slated for all eternity affect one's thinking in the short run? Is there a possibility of transfer between Hell and Heaven? As I asked no-one, I got no answer—but I did get delivered to my PO Box a huge bag of guilt, the largest I ever saw, and an attached message that I was to spend eternity by doing penance.

I view the indeterminacy of my prospects as my fate—not exactly what I wanted, I first wanted to be liked by all—but the latitude it gave was enough to free me for the growing certainties of my later time. I take it that indeterminacy is a kind of neutral realm which does not welcome certainty as a member. Neither the early scolds nor later syncophants—especially not the sniffers into my asymmetric virtues—those with voices that chant most shrilly when they would peer at my full-frontal vices—they are no longer welcome in my life. I am beginning (just now, imagine) to arrive at my own truce with pain and pleasure.

Late Pain

Eventually, we reach the cocoon of the pain within which we die. By that end-time, pain has shed its partiality. It does not pretend a further intercourse with pleasure. It no longer has shape or limits; it does not respond to either purpose or status, and it is not moved by past friendships or present solitude. This pain is immediate, pervasive, and everlasting—a proper introduction to hell—if not to heaven. This pain no longer teaches. We cannot, except in fiction, be so stoic as to tell it "as it is" to those caring pain-free others—the ones who stay too long beside our bed and look at us so curiously. They—they are sure—will not, not in their own time, be as they see me now—this wretch of smelly fluids and wasted limbs—peering into the mask of nothing much—nothing at all, really, as cognition dims.

The insistence of final pain brings ending into the obdurate and eternal present—it is like the garrulous guest who persists in drinking your best cognac and talks about himself after the party has ended. (How can I get him to leave?) All this bother from outside weakens the assurances of the secret self—the one who lives, still alert, still young, inside the dying one. This is the self who is sure (education has that virtue) that there is a way of thinking past (rather than just living through) such a thing as dying. Thought is powerful, but end-pain is purely somatic; it is is indifferent to thought. It is the fire that consumes large gouts of time and leaves only a trace of ash—which blows past the enduring stuff of memory. End-pain has no interest in the summing-up—the exhortation we all should make to the world before we leave: "Listen to me, I want to tell you what you now must know and what you no longer need believe." Perhaps the pain-god knows better: There is no exhortation, coming from a life however lived, that has any bearing on death's present. End-pain also does not care about death's future.

Later, a little later, but too soon for life's demand for a second try, the end of breathing is what those still around will notice. Then, the nice fat nurse, older now, comes by with her stethoscope, dangling like a crucifix between those wondrous breasts, and confirms to others what I can no longer know.

Whatever our separate beliefs, there is enough pain for all. In the days of my first dying, some old friends came by. (I remember them well—they are survivors of the early grubniks who long ago joined me at the San Remo for evenings of exemplifying "cool"). I know why they are here. They want to revive the past through the old-fashioned way by zonking me and changing my solitary hurts into a collective fantasy: "Listen, moribundo," they say. "What you are thinking now is what we need for the seventh chapter of our manifesto—which concerns what is still not being done about what cannot be undone. So stay alive for a bit and help us out." I reply in kind: "I will gladly speak to the undone of necessary doing if you get me high enough. As a first royalty, tell the sweet fat nurse with, you know, those breasts, to give me some major nipple. Tell her (before I have relinquished speech for sucking) that she need not fear—that I know the difference between enhancement and distortion. And I have some saliva left."

Other people, especially the old sequestered folks who never got to the San Remo, have problems with narcotics: "What will his last thoughts be if he is high as a kite?" "Will she still recognize her kith and kin if, God only knows, she is wallowing in some queer and dirty fantasy?" "After all, is being out-of-mind a good way to end a proper life?" But the dying person-in-pain can also be an irritation—there is little about the obdurate presence of the almost-dead that can be appreciated. The friends and relatives, who are summoned to the bedside, will ask of anyone who vocationally listens to the sound of dying: "What can we do—now that we are all together at the ending?" "Nothing now," the white-coats say. "Before was a better time."

Back outside in the parking lot, the witnesses will work to set the record straight: "Was he the uncle on your father's side, or the cousin on your mother's sister's side?" After that, they will talk among themselves about the comfort—and indeed, example— that old Moe, by dying as he did (never mind his last farts and gurgling grunts) has given to all of us who seek the way to a proper end. They will call it a public example of true-belief and a personal display of courage—"by god, God, he did it right." The family will describe the dying to those who were unable to visit as "blessedly brief; he didn't seem to suffer—and in fact, our presence helped ease his passage."

But pain, real pain, breaks through all this. Then we stammer, wave our hands, kick our feet beyond their strength—and cry canticles of "argaharananagrinawanamama." This is when the flesh constricts around the standing hairs—when the eye skews inward and no longer sees what remains outside as its lids close to contain what is left of the soothing dark.

Getting High

Yet, for the misbegotten and unrepentent dying, there are old friends—if they hear the news—who can provide the way out: Old time panaceas—opium and its modern cousins, shrooms, peyote, high-end pot, even LSD—will do the skippy-trick.

I have been told that afflicted souls when sufficiently high see their pain as a boat running rapids in a distant river, just barely visible from the clouds where they now recline. They watch the journey for a while, and then when the pain-boat goes beyond the bend, they close the clouds until everything is white and dense—obliterating even the last small sense of movement.

Others, perhaps with different needs, have described the time they became pain-free as being in the center of a herd of rutting elephants, where they had been invited despite their small size, to contribute their bit of exudate to the gobs of semen that flow into the darkness of the primordial grotto—there to be transformed into the pipperipping joy of a baby elephant's birth.

Still others, once themselves ambitious, recommend that when you know you're dying you should draw your feet—the ones that stick out from under the blanket that covers your too-short hospital bed. You don't ordinarily look at your feet—but this time, considering their long toenails, dry skin, blemishes and staccato bones, you should pay your own feet the attention you once gave to, say, Marianna's thigh: Draw, erase, and redraw until you get it right. Feet are not small; they are a landscape of rough high places and low fetid copses. You can walk the pen between your gnarly toes, crawl over the hard-working ball—slither down slowly while wielding your tickly quill across the arch—and then climb up onto the horny heel, where the callouses of a long–lived life are embedded.

None of these efforts at "dying as you wish" come easy—the nay-sayers, scolds, and pointy-nosed inquisitors are always around. Now that you are weaker than you were, they can press you with objections to your floating world: Consider, they say, before the perils of judgment come—your deviations from scripture, decorum, and family-values that have clouded the ending of a life—as evidenced by your mauling the buttocks of cheer-leaders, gimping after aging models, and way-laying other wantons you convinced to stay put rather than run. Yes, face it, you have lived a deplorable life—one which has primarily been devoted to sins of the flesh and sneak attacks against the righteous. We can only understand this scurrility as an infernal barter—a giving of your earthly soul to the Devil in exchange for the beastly joys which he says he can offer you in hell.

I don't deny it. Hell is a place of elision and ambivalence—just the kind of place I spend my vacations at. But now that you accuse me from the pulpit and on the internet, I must reply: I say to you—you jowly child-molesters—that it is strange for the executor of eternal torment, the Devil you so despise, to be concerned with easing earthly pain just when it begins to hurt real-bad—while the godly bear grim witness to the benefits of its getting worse. Pain, as you claim, has a long history of supporting true-belief. But think hard, you creeperoonies, when you advocate such strictures, about what it is that really ticks you off. Is it your own hatred of yourselves for not feeling enough of pleasure while living—having to sit astride your personal Devil while calling on your distant God? And why did you always, from your earliest days, find sin in each of your errant twinges—and yet later, gladly suffer twinges where-ere you tread? You don't admit that you do twinges? You say, instead, that your displeasure with my end-of-life is justified by the fact that, unlike you, I have acted on every twinge that came along. Mostly true—I admit it. But you also say—and here's the rub and scrape—that I haven't suffered enough—not as much as you will want to suffer come your turn. You must believe that pain is the essential part of that compact we all make for the originary gift of life and the entrance to the reality of the after-life—which is modelled on the howling rapture of the Crucifixion. You also believe, do you not, that I have escaped all that by titillating the Devil with my earthly talents. Quite true. But can it be, then—as my art has made the Devil smile—that not only pain, but also pleasure, are proper-parts of Hell?

In the low-lands, getting high while dying is not for those who believe in a rational outcome for a life well-lived. Being high is a wet and fuzzy affair that lessens hurting when hurting matters too much for tending—as one could once do—to one's life. The state of "being high" spans the earth-spattered time between last living and first dying by permitting a pain-free passage across the chasm between "knowing-that" and "not-knowing." What we solitaries, when we are finally caught in the trap of their lying and our dying, may want to say—when we can no longer sing but still have a tune in mind—is not what we thought we would say when we were still some time away—rutting and snorting in the labyrinths of middle age. But now we say: "The pain that hurts so bad is no way to celebrate the end of me—so who among you flatheads (my old-time buddies) is going to give me a fix? I have told you, as I promised you, everything I know about what must be done about what cannot be done.

The good fortune of dying without pain is that it provides a last look at the fading of our thinking—which is not a bad way to go. We watch the room darken and we begin to think in circles even as the sun continues to follow the straight line to its setting.

Dying in pain is not a fitting end to even the least of us. Like the setting sun, it follows a straight line before it reaches its expected end. Circles, in contrast, are better—they alternate bad days with good. The whirling can be fast or slow, but speeds of whirling circles are hard to determine—for, however hard you look, you cannot tell if they are moving or at rest. So coupling speed with achievement, progress, and salvation, produces a straight-line image—which blurs when the snake gives up the hunt and turns around to bite its tail. Circles show us other ways that we—and the world—could come and go.

When I go, I will want a fix—not an ordinary high, but one that pilfers my past for the more hidden episodes which now can be scattered in the Dada way—and so show me how advanced I really was. My constellations, I find, have good form and lots of juice! Having now reached the end of my time, I no longer need to woo the latest reality, but can dream of other ways to go a'dying without mocking others' separate quiddities. I am now, as I want to be, surrounded by the creatures of Christmas Past—as magnified by (my) world-memory and free at last of (someone else's) irony.

But if the fix is not there—if I have misinterpreted my rewards for having lived at all—then I must crawl into the sack of ordinary suffocation where non-breathing is a first transition to the next state that really matters. Non-breathing, when compared to what breathing demands of us, is an option that can be taken on when the in-and-out of pain refuses to stop.

In any case, the raptures of the end-game should not be circumscribed by intensity but rather by completeness—for dying is all about completing; it is about the completion of a life—not of its lived ambitions. Intensity is about beginnings—prior to ambition—such as when being circumcised is an early step to recognizing the fact of life and pain. Intensity is a transient gift to consciousness but it is important in its consequences—for who would want the cutting to be soiled by some remnant of foreskin that was left to dangle into puberty and beyond? The end-game is different—it has spoilers more complex than a simple slip in slicing to subvert its rules: Among the hazards of dying are unfinished tasks and unattained ambitions. Dilemmas that afflict the dying are self-inflicted accusations—a distillation of years of failure without sufficient documentation of the occasional success. In the hospice-quiet of self-accusations, they can become guilt-trips donated by you to those you leave behind. Don't do that!

But do not despair, you moribunders, tear-jerkers, appointed-mourners, caretakers of the customary rituals of dying. I say that a deeply-lived life can be enjoyed while dying if its memories are thick enough. But that depends on how you had learned to live—how you balanced memories against ambitions. Before your life is over—before it comes to be your nick of time—an estimate on your behalf can best be made by, yes, you. This, your dying, is the last time you will have to grapple with the accusatory beast who, since childhood, has given you such trouble. Now and forever, you can kick the bastard in the balls. To die while shedding guilt and enjoying what you know, is a good thing.

Dementia

Then there is the dying before you're dead. There are many medical terms for this: Dementia, Parkinsons, Altzheimer's, Lou-Gehrig's—the list goes on. In all of these, the brain is attacked by substances that progressively destroy the neurological sites that normally link our cognitive capacity to structure the world as presented to us by memory and sense. In this condition, pain and pleasure both become irritants, and practical decisions are no longer thinkable. However it goes, fast or slow, telling becomes more difficult: Responding to the recognition of a voice, reacting to a caress, eating the spoonful of diluted food, moving in response to the sponging of your genitals, even acknowledging a pin-prick, slap or squeeze—all these diminish and slowly disappear.

The supposition is that so afflicted we think less and respond more feebly, until we are without gratitude or anger, joy or sorrow—and we are reduced to sitting in a chair where even our daily anguish at the dying of the light will disappear. Then we face the outside with an unblinking stare for as long as it takes. This is what others see when they look at the wreck of us. But they don't really know what is happening in the inside, in the mind of the changing brain behind the mask—and we can't tell.

Arthur Danto, writing about deKooning's "Altzheimer Paintings" sought to rescue them for art-history by questioning our ability to understand the workings of such afflicted minds—and our concomitant willingness to put a negative value on these last works. Certainly, the works on-face are more complex than, say, the recently heralded blank canvas, or the artful splotch of color on the empty gallery wall. But deKooning couldn't talk about his last works—and few others knew what to say by then—because the works were so separated from his earlier efforts—and he had become so inscrutably ill.

Yet, there is the possibility that a valuable (however defined) internal and (for now) unreachable process is going on in that unresponsive incontinent hulk before us. This has not been fathomed by neurologists, who want the brain to exhibit patterns of behavior that show (that are causal to, not merely coincident with) what the mind thinks and knows. But showing what one thinks is also a historical and philosophical question: What is the mode that we recognize as showing thinking? Is there something being thought that is not shown—as far as we can tell? There is a whiff of predestination here—a mode of prediction which catalogues a life through its external manifestations—bumps on the cranium. But life goes inwards too. If the prediction does not come about, if the paintings do not pass the test of intentionality because of the lack of access to that accursed free will—embedded in that, perhaps, non-existent mind—then we should obey our hunches, buy some now, and wait for a revision in the criteria for aesthetic value. Then, it might not matter what the afflicted artist meant.

How does one find out what is going on inside a reclusive consciousness? Medical research goes as it goes. But there is a relationship between social needs and research emphasis. Comfortable societies are negatively sensitive to aberration—it threatens stability. Dementia is indeed an aberration—but it is the image of dementia that is most damaging—to caregivers most directly, and by extension to the larger society. A wasting body that forsakes its own functions, and a mind that is manifest—if at all—in inaccessible ways is no longer human. This goes against everything we think of as a valuable life. It is evidence of something more mysterious—a withdrawal of consciousness from the world while still living. The mystery lies in where it (consciousness) goes, and how far we can follow it. Of course, we need not assume that there is such a thing as consciousness; and we can use Occam's razor once again—to the impoverishment of our own psyches and our ability to empathize—however irrelevant these may now be. Or, instead, we can pursue the unspeakable and unthinkable into other worlds—however foreign these may still be.

The haplessness of pain is mirrored by (but is not the same as) the nullity of dementia—although both show an extremis in their necessity. Pain separates choice from sensation—exterior and interior of the sufferer mirror each other through their spasms of avoidance. Pain without hope is worse than encroaching nullity.

Dementia also shows what looks like decreasing choice, but it may only be the inability to react to, and articulate, external sensation. Dementia supposes a counterpoint between two mirrors—one reflecting out, the other in. The first, from all reports, grows dimmer as the disease progresses. But the second might well grow brighter by reflecting an internal world that loses dependency on (or compensates for the loss of) its external counterpart by becoming incrementally composed of the figments of an increasingly disembodied memory. Such a memory, given its occluded structure, does not participate in the selectivity and the regimentation of the practical memory—which is parsed by interactions with the world that we engage.

But this deprived memory, sequestered as it is from external stimuli, might well have a source—better, a storehouse—of other stimuli, namely, the unregimented memories of all experience from its own past—memories both remembered and resurrected. For those who have access to the world outside them, memory is a matter of consonance with present experience—but the store of memories is not limited by consonance—its whole is larger than its function in the experiences of any given time or place. In this sense, memories are inexhaustible, although they are not infinite—they become a finite (yet unreachable) set the moment we die. But memories need not attach only to experience; they can be consonant with each other—even in the absence of external imput. We can dream about dreaming about dreaming… Recapitulations of times past, on this reading, extend each other and so can provide a separate world of images—although they aggregate in sequences that we, when wide awake, are not privy to—unless, of course, our access to the outside world (as in dementia) has been cut off. Then they become our only world.

From another, even more solipsistic vantage: The movement between outside and inside may be just a variable in consciousness—an outward shift that responds to the fecundity of external stimuli of profit and pleasure, or it can be an inward shift as a reaction to threat or deprivation. Outward shifts occur as actions whose consequences we recognize, but whose reasons are only traceable through introspection, and whose origins may be other than the reasons so revealed. About these origins, no-one out there nor in here can know—for to know them requires a deeper diving than most of us can do. Perhaps it requires an unusual willingness to stay with what the deepest divers find below. Perhaps this is the condition of dementia.

Inwardness can become a preferred state if it succeeds in repressing the external conflicts of excessive confrontation, undue triviality, or unruly nerve-endings. When the travails of externality are too-much suffered, they stimulate the turn to inwardness in a time of life—that time when one should be most bent upon attaining wisdom (or, at least, tranquility) before one dies. However, if all laid-down reserves are depleted, the only recourse is escape—down into the neglected swamps and bogs—pest-ridden though they may be—of suppressed or forgotten memories.

Memories

Pain and pleasure do not offer an equal contrast for memory. Although they both situate in the construction of past and present, they vie at every stage for recognition. Sometimes, we do remember pain, and find the memories intermix with memories of pleasure. So we scurry back to knowing ourselves by saying that in any actual state of affairs there is no pleasure without (some) pain, and no pain without (a little bit of) pleasure. This shows our preference for moderation. Pain, however, being stronger and more direct, can expand to where it shunts this preference aside, and so becomes unbearable. Pleasure, more ambivalent about mission, continues to seek—however enjoyable its own ascendancy—some residue of its other; pleasure never becomes so pure as to be unbearable.

Sometimes we cannot tell the difference between pain and pleasure. Sometimes we cannot "know" them—which seems a requisite for telling, but we can "feel" them—about which feeling no-one else can know but which seems sufficient for our "having" them. Without the knowing, we do not have the grounds to tell our preferences—without the feelings, we would have no preferences.

All is not lost however: Distinctions swim by even in the fish-pond of memory—the early stuff that feeds our later life: We remember the fright in our first moment alone on the potty, and trembling in our first nakedness before a stranger, we continue to relive the assault on nasal passages of a first swig of hard liquor, and we return (daily? weekly?) to that heavily thumbed array of early drawings of girlfriends.

Sexual role-play is an ever-popular subject for fantasy—whether encountered in dreams, projected through media, seen in theatre settings witnessed by uneasy but attentive audiences—or played out privately at home. These variants are especially captivating when they appear as memories of our young and reticent mix of pain and pleasure, and then translated into the latest styles of titter howl and whimper. Such memories are our early creatures—which we once could not have admitted to even having—but are now enacted, out there, for all to see, by players who are as proficient as we at slipping between fact and fiction. Sexual partners offer themselves to us, but rarely offer us their dreams. Our "having" them, then, is incomplete—what one calls "metaphorical"—as is equally true of the "offering."

There are memories I do not offer but especially cherish—of salty hors d'oeuvres, smoky rooms and ice tinkling in pitchers next to glassy eyes and slipping shoulder straps; they take me, as the night continues, to those dark places where, as I remember, we do it over and again into the morning. Indeed, these may only be memories of my dreams, and I wonder now, whether I mind not knowing if they really happened—if they are more dreams than memories. But my cherished memories are not restricted to such things as belly-buttons and the sweet-sour smell of bottoms; they are interspersed with fresh straw-berries, slow-smoked barbecue, and painted wiggly toes. So they must be mostly dreams.

But there is also the pleasure in dreaming—whatever else you remember—that you have lived life well. There were of course, alas, the many times of lying and prevarication, sloth and indolence, carnality and spilling on the ground—all the sinful things you push off into forgetfulness even when you no longer believe in sin. Further back in memory such paths are dangerous to take, for they lead to the ancient painful places. There were the times when, still believing in sin, I felt the eyes of the Virgin (or was it the Magdalen?) upon me.. Those implacable eyes found me most defenseless at night—when I was cold and shaking and wetting down the sheets. In truth, I was more drawn to the Magdalen than to Mary. She was fierce, yes, but I thought that she could teach me more—because she had done these things. Mary, they say, knows nothing of such things.

I haven't, in any of my lives, remembered everything, or even every part of anything, that happened to me. My memory, I tell myself, should contain only the things that fit my story of the life I live. Of course, there are the unbidden memories—but these are jagged, oversize, and out of focus—coming mostly in the wee hours of morning just before I have to pee. Then, duly awake and empty, I try to suppress the sporadic content of dreaming with the linearity of waking life. But some dreams, like nightmares about the fear of fucking, can be intrusive, and will stay around and invade the waking world. As I got older, I could shoo such dreams away, cast them back into the river-bottom where my friends and I no longer want to wade—especially because upstart minnows now nibble at our toes.

Sometimes, though, unwanted dreams have hidden memories that are so fierce and insistent as to go around the back and attack our waking defenses. And they sometimes are victorious, invading the sanctuary that we, in good faith, have erected for reasons of not wanting them around. Then we lose the hard won equanimity between the images of our present and the memories of our past. Sometimes though, we see a memory—our once private gargoyle—perched on the shoulders of someone else. Then, to prove that it is ours as well, we let all the others out—a trickle become a torrent—and insist that everyone look and listen. That's a good way to become an artist.

A Way to Go

The guidance counselor said to young Lucian: You are high on social skills (although your ghetto origins do not bear this out) but you are low on computational skills—so no physics or chemistry for you—don't try to be a doctor or a banker. Although you seem inordinately drawn to fantasy, poetry, consorting with bohemians and badmouthing nice folks like me, you do belong in a university—especially a public one like ours. It is, after all, the floating barque of your salvation—full of older creeps who once were just like you. When you graduate, find a way to say nice things about professors. Right now, look for classes you won't fail.

I found that in art-classes I couldn't fail—we made up what we had to do as we went along. It might have been different were I an apprentice in Titian's studio. He might at times have beaten on me with his largest brush—but I think I would have made it there as well—learning the pictorial skills to paint his landscapes and, eventually, the drapery. The nudes, of course, are something else. His projection of prurience was far more inclusive than mine. He could travel from earlobe to armpit, and from there through crotch to ankle and have us believe that we were circumnavigating the globe.

But I am old now and seldom travel, and I am content if you view the demesnes of my small place with large expectations. In good light you will see the many ways. The troops I have enlisted to move my goods fly many flags, cunningly configured like the banners of medieval Polish cavalry—to make the few seem as legion. This is in keeping with my ambition to make an art that shows the differences a life can have (mine as well as others). Fly enough flags, I say—even ones not entirely of your making—many flags will win the day—maybe even the goddamn war!

I expect that you have met such poly-enderers before. But don't go away. In the transparency of the midday sun—in the light shining on an optimistic art-world—informed polytrists do well. You can grumble that you are dismayed at some images—gnarled and bulbous forms stumbling back from the classic battlements they had once tried to breach—they don't even remember what their anger was about. I tell them to try on other clothes. Now see how pretty they are coming back upon the ramparts—waving their banners of the new accomodation. They are what the evolution of spirit has in mind for me.

We now know that the classic smells—the fresh sweat and grime of Socrates' youths wrestling with him in the forum—would eventually be cleansed when artists came to insist that to be true to life, their models must taste and smell in the same way they are portrayed. Friend, breathe deeply with me while you sit on a cushion in the middle of Ingre's "Bain Turque." Do you not find some scents that are familiar? Try breathing deeper—perhaps a bit higher or lower. Smell, as it is more immediate than sight, gives assurance that its presence will point to all the other sensations that we would-be Turks need for a life. Now that you are practiced, can you find a place—a way—between the thighs of the Grand Odalisque that would woo her into assuming an unaccustomed awkwardness—a mismeasure of legs—a sag of breasts—harbingers for the art to come?

During the invasions, we Barbarians did not want, on pain of gross confusion, to talk with victims about their smells. We were content to simply fuck before we killed. What could smell tell us that would further our cause? But it came to be, in the early flowerings of ecumenicism, that men put aside both power and status, and faced the women on more level ground: "What do you use to clip your toenails, we asked—myself, I use my sword. Does pubic hair in your town get cut or plucked—or does it just grow wild? You can use my sword to trim yours if you like. Is the practice of smearing sheep fat on your wrinkles a mark of privilege or is it available to all? I'd like to try it. Do you have a personal trainer? I don't either."

Pain and Pleasure in Art

It is hard to write about past pain; one remembers only that there was pain. How can one be true to writing pain? There are the familiar locutions: "Agonizing shrieks;" "Wracked limbs;" "Piteous pleadings." These are all Gothic fantasies replete with pornographic possibilities. But when you try to say it straight, the words begin to shed their sense like the skins of last year's snakes—whose subjects emerge pristinely new, and if only for a short time do not look for a better fit.

Writings fail to approach the abstract beauty of the Crucifixion—a central subject for the pain-obsessed contingent of western art. What has been written about the event, from the gospels on, is a barrage of testimonials, exhortations, history, homily, explication, catechism and political mandate—but very little concreteness about the pain of bodies. Augustine and Luther gave reasons for our believing that they have conceptual pain; they told us how it is to have it, and they tried to tell us what to do with it.

About bodies, the painted versions have it better than the writings. Last Judgments—when made at the height of skill and belief—as in Bosch and Giotto, Van Eyck and Michelangelo—profoundly picture the tension between the saved and damned—pleasure and pain—even to the modern eye. Now try this: Line up the great crucifixions painted between, say, 1300 and 1800, and you have the history of the underlying attitudes about pain and its necessity, pleasure and its pitfalls. Go on a pilgrimage to where the paintings are. They are mostly in Northern Italy, although the Germans and Flemish have their good share. These visions are the center of Western art.

In music, Bach and Handel captured the spirit of belief but not the pain. The Messiah, or the Saint Matthew Passion, evoke our awe that there are beliefs so strong that we, whether believers or non-believers, are carried along their revelatory path. The revelation of pain in music comes later—where it shows a conceptual emptiness in which pain is a consequence of irrational or indifferent forces. Schumann's "Dichterliebe," and Schubert's "Winterreise" reveal the pain of individual alienation and impotence—the fate of the romantic poet. Schoenberg's opera "Aaron und Moses," is an ancient parable of inescapable fate—but it is given a modern content through its dodecaphonic demands on appreciation. Berg's "Wozzeck "—arguably the greatest of 20th century operas—gives expression to the "everyman" caught in a vise of social and sexual demands he cannot meet or even comprehend.

These days, pain is readily captured without such trappings—on film and video. There is ample footage of every-day horrendous practices that need give no recourse to the victims. This is a particular issue in documentaries: When does the film-maker stop shooting and interfere in a circumstance that is morally unacceptable? When does the artist sitting on the boardwalk stop sketching the drowning swimmer and try to save her?

But where the mode is fiction, the representations of pain or pleasure are diluted by disbelief: What you see, however convincing, is only a late varient of our early games of make-believe—cops and robbers, cowboys and indians—but, as I remember, despite our assigned roles, we only played us. However, the multiple selves required of good actors goes beyond self-conscious "play-acting." A role, say, of a victim of religious persecution may follow a previous role of a Victorian dandy. These differences do not hinder the audiences from engaging vicariously with each representation. (Think of Meryl Streep as Sophie and as Margaret Thatcher—or Charles Laughton as Quasimodo and Captain Bligh). Stars are particularly good at filmic suffering. They can mount some mighty dreadful howls and we will gleefully shiver at the fine portrayal, knowing all the while that John Wayne will leave his present fate behind to return, in a half-year or so, for a different attack against the overwhelming powers of evil.

Other Occasions

But you know all this, my poor suffering soul of a friend. You know that these images are too histrionically enlarged to offer any help for public viewing of your actual back-pains. Of course, no one wants you to have them, but certainly, no-one wants to see you having them—even on film. They might be real—the only part you play—one never knows these days. Besides, no one out there has thought of anything that would celebrate the greater implications of your pain—and so spark interest in the film industry. Nor will anyone come to make sketches, or take some photos of your sufferings. Old soul, you must admit it: You did nothing especially right or particularly wrong to merit such attention.

But even so, you didn't take the chances offered you. What about the pain you showed last Christmas—when you upset the punch-bowl with your thrashings, and tried to wrestle cousin Betty to the floor? They're all gone you see—you didn't follow up. Punch-bowl's broken and Betty's dead (of natural causes, to be sure) and nobody was there to film it. You don't even remember that time well enough to say much of anything to anyone about it– not enough to make others understand how really important the slippery floor, the broken glass, Betty's veiny thighs, were to you—why, they are the very stuff of filmic fiction. You needed Ingmar Bergmann to capture your ongoing contortions. But let's face it; his interest would only be to use you as a minor subject, standing by the outside door of Sodom, marvelling at the central enactment of divinity's concrete manifestations of pain.

Your own pain, by all accounts, does not compare in poignancy or substance with the sufferings of the major sinners. So you don't need old-master attention right now—how about a porno-movie, or a cup of tea? No, you can't have any more wine—the doctor said it's bad for you.

Let's change the subject, and look again at pleasure: Ah, we see them there—the ancient memories of youthful ferment—now ready for drinking. We bubble up the images of past ecstasies with every glass—yet we assure our friends that, although those folks were faster, we are better off now than at any past time. Slower is better, we say. We lie of course.

It might be nice, though, to have another go (even now) at breaking through the shifting and obfuscating fog that surrounds past pleasures. Pleasures do well anywhere, but tradition places their best fecundities within a fog that obscures sight but not sound. Walking in a foggy lowland and hearing screams of rapture whose origins you cannot see, remains a preferential memory—much more than do howls of pain. With age, it becomes easier to tell the difference. Our memory, through which we now say we can dispel any fog, is not so much for our delectation as it is to show the way (an old-age gambit) towards diminishing the act—the dirty doing—and embracing the form—the Platonic Ideal—of pleasure.

Strange, though, about pain and pleasure: Another's pain evokes public sympathy, covert satisfaction, and increasing distance—but another's pleasure evokes envy at not having been asked to the party even when all those other losers were invited. How then, in the continuing world order, can pain be ameliorated by pleasure if their juncture is always marred by reference to conflict? Memories of our early pleasures should be guides for our pleasure-seeking in the future. But this can deceive. We know that our past pleasures were mostly stumbled on and are not repeatable. The principals have gone elsewhere and, for all we know, do not want to remember what once happened in that dead-of-winter, fifth-floor-walkup, cold-water-pad. They should remember—it was nice.

But those bright and shining ones, for all I know, are all dead now. Also, I am older, and the lure of resurrecting moments in past life to include them in the present of my daily meditations, is less appealing than the understanding that things don't work that way: The present is not eternal; memories are not reliable; both pain and pleasure have an end.

It is difficult to compose a text that tells the story of pain or pleasure in a way that can be shared. Why, anyway, should you, dear reader, want to share my pain? Nor could you—however much you try getting past my ever-watchful dog—enjoy my pleasures as I do. But suppose, anyway, that my literary efforts, my descriptions of pain and pleasure, should overwhelm your reticence and have you accept my invitation to spend the evening hanging on my cross of words, or squealing and shuddering as you are stretched on my didactic couch. After a morning shower, though, the face you look at in the mirror will not show the same concerns as mine. What was that about—all that shouting and singing, scampering on all fours up and down the dirty stairs, and swinging naked from the balustrade? When you are clean and dry and breakfasted you will try to forget those images. You will want to know, in practical terms, the difference between what we rhapsodize about, and what one feels when one feels pain or pleasure.

Pain, whether present or recurrent, does not hone a sensibility or a body; it only makes the victim empty, skinny and twisted. But such pain, just because it seems intolerable, is always blamed on something, anything—the inheritance of malevolent genes; the casual brutality of muggers; the purposeful infliction by inquisitors; the glee of perverts who also like to watch, the sickness unto death.

In contrast, pleasure makes our bodies fat and our minds smug. We describe our pleasures as proper or hidden, too fast or slow, too intense or barely there—or a bit too pricey. But, unlike with pain, we are cavalier about the vagaries of pleasure, preferring but not insisting that they be here or there or more or less. The pleasures of the flesh and the delectations of the mind, despite ascetic grumblings, are more similar to each other than they are to pain. The form of pleasure's presentation depends largely on the habits of your fellow voluptuaries—whether they prefer to taste the chateaubriand before they settle down on the connubial couch, or pleading indigestion, flee with unmannerly haste to the safety of the street.

But I must tell you—there is nothing more pleasing to the mind than the rapt co-ed gazing at the tweedy professor as he expounds on the mysteries of nothing in action. Now, that is a titillation experienced only by the privileged few—but difficult to find these days. (Co-eds, these days, do not read Heiddeger nor can they bear the scratch or smell of tweeds—and the savory juices of a well-aged steak will kill you). Our present pleasures tend to follow the blandishments that govern mass consumption, and so, like the latest model cars, encourage competition between versions of the same dish. Serious feasting is hard to sustain after all the comparative tastings.

Pain, being solitary and immediate, is why the divines still prefer it as the soul's pathway to heaven. It suits the reclusive artist who is pulled despite all resistance to the river's edge by the hungry crocodile. There he can watch his limbs devoured one by one—a truly worthy subject for his talents—until the fearsome teeth reach some central place at which there is only time for one quick poem before the dark, with its unseemly haste, covers all. For the beast, time slows, as it always does, to meet the needs of digestion— an artist is little different though less tasty than a young doe. Blessed are the eaten, for they travel light to the kingdom of heaven.

Pleasure, being concerned with the compliant intact body, goes in other directions—it first arises from memories of the last person you enjoyed—and through collage, brings back the earlier memories you were smart enough to splice and save. The glassy eyes and screechy yowls coming from fully limbed companions in this evening's bed, are given resonance by remembered vignettes all come back for a leisurely time of it. I sometimes want to give each such performance its separate due—but that's so back there and then. We now all recognise the sameness of terminal sensation—just the heft and colors are different. Even so, the players in my fantasy should have some say about how they are portrayed, if even only in my memories—but that surely isn't now. Oldies, such as I, who no longer look for ecstatic coupling, will want only to sit demurely by with their martinis.

But face it—our memories tend to screw things up for each age we are as we get older. They become less seemly and less tasty. It would I think be wise, when having reviewed our recollections– whether of pleasure or of pain—that we deny others access to the episodes we like best. Those are the episodes in which pain and pleasure mix, where we recognize the increase in erotic value through an emulsion of sensations. We must retain them for our own identity. They are not for others to know.

Variations

But what is it that ticks you off—venerable father? Why do you so dislike those who choose not to acknowledge your presence in their old memories of young prurience? Is it because they insist on equal access for all those unsummoned and unconsummated ones—the pretty ones you barely knew—and the fast ones who barely knew you? They all once were somewhere there, with those wisps of thigh, softly curving breasts, and firm beliefs about the way of flesh—which gave them, Marxists and Baptists alike, a way out.

It then seemed very rational, very free—but this freedom had a built-in imperative; it always insisted on change—and before I could get the sequence right, they were off to find another who embodied a more comprehensive, faster, more forward-looking freedom. That insistence on "getting it right" was very painful for me—it encouraged my loves to live in someone else's memory. But other loves did leave others' memories to come across to me.

Oh, they will now protest, these ancient ingénues, that the life in memories is just fine. And, given past particulars, (the lapses on both sides) why shouldn't they and you and I again be the stalwarts forever fornicating in the make-believe theatre of the everlasting? But those early memories (there was no video then) are now showing rust. The reedy voices of past years, who insist that we again be like what they always wanted us to be—they lie. They only want to live their inheritance within a clean and cozy condominium on the upper west side—walking distance to Zabar's—and they no longer care whether carnal pleasure is receding by the month—what beauty looks like—or why the days continue to be so much like each other.

To counter the uncertain realities of memory, we must turn to the clarity of fiction. Think, if you will, of the Classic pose—the stretched articulation of the limbs—that Poussin gave to his martyred saint when the good soul's entrails were being rolled upon a flaming spit. That's a powerful beauty dad—and more politic with its beguilements than poor old Riemenschneider could venture—he who tried to carve it as it really is and had his hands cut off for his efforts. Only the Devil could make a statue be so realistic, they said—yet he continued carving, with his tools tied to his stumps. I admire this, but I would flee these northern monstrosities for a more circumspect safety in the Southern climes, and so I recommend Poussin's way: Give pain to classic beauty in the way it hurts its long-dead subject—and delights the rest of us.

But this is an ideal that commemorates a vanished world, and we live in the present world. In this our world, there is no beauty—however much it is entwined with pleasure—that is without some lurking pain—some memory in which an expected joy is not fulfilled because of the appearance of another, distaff, memory. Consider Titian's "Venus and the Lute Player," which denies to any mortal the pleasure that the lute-player, had the gods made him more priapic, would have taken for his own by transferring plucked melody into the sounds of fingered love. But the gods didn't—and he couldn't—so the ideal source of pleasure in the western world remains somnambulating on her couch, indifferent to the potency of her image as a model for our desires.

So I ask you: What would you say to Venus if you had by chance been invited by some subversive minor god into her chamber—when the greater gods were away? I know what I would say: "Venus lady, baby—if I may so appellate you—I want to admire your digs and kiss your dugs and smell the perfume of the flowers that surround your bower—and later, if you don't mind, I want to sip the nectar that wets the common-ground that is the meeting-place of gods and mortals. But get that musician out of here."

Washing is a habit that is periodically indulged in on earth, but is not necessary in heaven—so Venus will not smell, I'm sure, the way Alicia or Bethany or Gertrude or Priscilla do. Now, I wouldn't mind smelling heavenly funk (when emanating from the goddess) even if she smelled worse (more ancient) than do her unwashed earthly counterparts. This is because, having in my day traveled some little distance between heaven and earth, I know that smells in both place are configured for potency (their empirical measure), context (their political role), and content (their individual appeal) as befits the diverse plans of both gods and humans.

In doing this, I have found many variants. There is, for example, the unwashed funk of winter months that only the most needy or the most lofty will enjoy. The needy bring the choking challenge of the smell together with the warm joys of rubbing under covers, and usually find there is no quarrel between the two in sustaining life and love. The lofty ones, when it comes time for them to satisfy their lower urgings, will call for attendants to wash and warm their subject, and so separate fine rubbing from coarse smelling. This restricts the privilege of being sniffed by gentry to the now pristine vessel who, in poorer circumstances, would not have had the luck to even be auditioned.

But there are places in the world where pain, by edict, is not diluted by the smell, or sight, or taste, or sound of beauty. The inhabitants of these places are arid, and neither cleansing waters nor fine emollients can moisten their flaky skin. They base their preference on a search for truth as it is witnessed by abstinence, obedience, and dry funk. These truth-obsessed are water free; they abhor the shining moments of the past, and take the wetness of enjoyment as evidence of false belief. The future, they say, is not the apple for your pie (remember what happened in Paradise) and is useful only when it confirms what has been forbidden in the past. They warn all who bathe: You had better, brother, dry and eat a different pie—one more humble— containing skins that give endurance to your coming certainty. The prickly stems and bitter seeds will shield you from the multi-colored conceits those clever devils dressed as wet and naked heathen want you to believe.

Degraded Pain

It is hard to accept that continental shifts, corroding sea-swept rocks, or parched grass-plains separated by intruding swamps, are candidates for pain and pleasure. But there undoubtedly was some point down the evolutionary line, where fishes, dinosaurs and birds, experienced something like how we understand these states today. Yet we hesitate to accept the equivalence. Some theorists call animal pain "degraded"—therefore not admitting, say, amoeba, salamanders, snakes, chickens, cows, whales, deer or even dogs, into our moral strictures against pain's infliction. This notion gets more problematic when faced with the pragmatics of slaughter-houses—and it becomes bewildering when Bambi is shot dead by red-eye and not even butchered for the family meal. But our not accepting animal pain as of a kind with human pain is likely to continue—the "sporting-hunt" is part of our evolution from eating to live to just enjoying the kill. But such enjoyment approaches a kind of theatre. Do we really prefer the death-rattle to the bacchanal? Is viewing an animal's pain more pleasurable than showing faked photographs of our hunter's stance?

The thesis that hopes to save us from our more unfeeling selves does not primarily rest on evolutionary or physiological data about pain intensity across species. It rests rather on the ideological claim that the lack of language in animals is a sufficient justification for how we use them. The claim goes this way: In our own children, the acquisition of language is at once evidence of their humanity and our imperative about how they should be treated. Language, here, is the faculty through which children progressively tell themselves and us how it is with them—and this places us all within the civilized cocoon (the meaning of life) in which growing pains transform into maturity when the protocol for this transition (family, education, wealth) is observed. For us, the protocol functions through langage and its translation of experience into culture and history.

There is a different transition for animals—closer to the moment of birth, less dependent on the extended family, more vague about time and obligation. Not having language, animals have no way to plead their case— even to their own. So they experience pain according to the happenstance of their own development. Their utility is to provide food and procreation for each other, and if they are of the right kind, to become clothing and food—and sometimes pleasure—for human consumption.

One must wonder why the practices of hunting and fishing—of family feasts around a boar run down by dogs, of shooting ducks and quail and spitting out the shot while chewing—have not been much criticized these days, when few seriously hunt for food. Although it might be fun, the numbers, admittedly, are small. In contrast, we breed untold tons of cows, pigs, sheep, fish and fowl, giving them life for the sole purpose of satisfying our need to eat or wear them. This is a different part of what we count as pleasure. We breed them in great numbers in order to kill them in great numbers for our greater need which, as we believe, outweighs their pain.

There is this other side: The six-pack hunters, let loose each hunting season, are more handy with killing than we consumers—despite the talkshows—are with cooking. Good hunters know how to pull the innards from a six-point buck; they know where the musk glands that must be cut away are located, they know how the saddle of the deer, the best eating, can be extracted out and the many ways to marinate it, and they know how long to hang the carcass for its role in the seasonal spread of merriment and feasting.

There was no further response asked of the buck beyond its startle and its bounding before it was shot. It stank mightily then, and attracted the worst of biting flies as it was being gutted. This was not the animal's revenge—nor was it taken as such by our week-end buddies. There was some talk, not much, about how the deer should be divided—comraderie ruled. Then after more cleaning and wrapping and a brew or two, the pickups started home to let the victorious hunters regale the wife and kids about the big-buck peril and the fusillade that dad had mounted to bring it down for the general pleasure.

XXI

Beginning, Middle, End

Pathways

The communion between the denizens of wake and sleep—between the interrupting sunlight and the interior illuminations of thought—as I remember it—was an easy passage when I was born. But it has been a difficult one to maintain. At one time, this interchange bridged my birth and the outside world. Later, it provided pathways—some steep and others slippery—to the hopes that others showed me, whose pursuit I expected would bring together the segments of my early life. Now another bridge is needed—between the continuous world and my increasing awareness of the fragment that I am.

I expect to walk that bridge a few more times—back and forth and back again—waving to dead friends and pointing to the spaces in the mean streets and musty buildings where we once were all together, exhibiting ourselves—writing, lecturing, painting, and yes, drinking and loving. They are the many I dedicate this to. But not really to all of them—I exclude the ones who tried to convince me that I should be just like them. My obeisance is to the others—innocents who shared my bed and the innocents who didn't.

I have visions of singing a song in winter when I reach the mid-point of that bridge: "In Questa Tomba Oscura, Lascia Mi Riposar." It is a sad and sentimental song, proper to a shapeless hulk in an oid parka and worn boots. On warmer days, I sing a summer song especially written for young girls whose thighs are shaped by flimsy dresses: "La Ci Darem La Mano".

But I must tell you, Signor Giovanni, you remind me a bit of Socrates. You are both seducers who follow your needs even when you know they will take you to death's darkness and perhaps the pains of hell. Cavaliere, if you had been wise you would have stayed with Elvira. She really loved you—for the pleasure of your blather as well as for your swift and smelly crotch.

In my early experiences, pain and pleasure were never the same in duration, intensity, or kind—but they could be seduced into forming a pact—despite the differences in sensation—which would give them a privileged, although unequal, status as co-conspirators in a specialized instruction for living: A little bit of this and a lot of that.

I later thought (more modestly) that sensations refer their differences to my brain-lobes via the data being fed by vast stretches of ganglia into those sensate organs (eyes and ears) that are their colonial outposts—and then into the brain/mind where they would be acknowledged as kind, quantity, and quality.

But I soon found that this description, true as it might be for certain sequestrations, was not always to my interests—which were more about desire than neurology. Yet, I could not clearly establish the boundaries of my discussion, and so its demesnes had to be drawn by other emissaries who—by ancient treaty—have access to the configuration of my ideas. This type of invasive creativity is, of course, not mine alone—but it does require an internal permission, a largesse, a letting go of empirical righteousness when the question calls for other ways.

The broad form of such largesse must have been drawn up a long time ago (Buddha showed great sympathy). Its performance gives power to such dignitaries as the God and the Devil—however fictional they may be—and it tests the beneficence of lesser, but more concrete, authority figures—parents, priests, teachers—all of whom, by mandate, have a say in what one thinks and does. It even (especially) gives power to those who have died.

Most of us, you know, will end up (if we do at all) in the unending grey and temperature-controlled expanse of Limbo. Once there, we will find, according to those who know, that we no longer are in need of unrequited passion or unrelenting pain as a way of understanding what and where we are. But eventually, however pleasant, we become bored with the timeless present, and we think back into time (nostalgia is endemic in Limbo). So we ask the reticent functionaries in this faceless realm of after-life, if they would allow us—just occasionally—given that eternity is so long—a life-reflecting gift. We will request some wine, and maybe a few tries at lurid and unspeakable practices—just once or twice a century, we say—perhaps some wildflowers, or some pornography in living color—as a memory of Life-before-Limbo before we again turn our faces to the forever place where spirit is neither dark nor light.

Long before, there were some who took it on themselves to instruct me on how to cope with my sensations—and their aftermath. Who were they? Why, the elders of my tribe. They clearly saw me as a prospect for their purposes—which they agreed should be mine as well. These notables are the ones (still) who theorize that it is incumbent on one not to say what should be said—and then they justify this by saying that "what should not be said"—taken in its widest historico-theological sense—is the same as "what cannot be said." They, of course, are wrong: "Cannot," unlike "should-not," is like the sea and the rocks—the first is beyond us, the second is up to us.

Others who were watching me—the more perceptive of the Heavenly Host—knew that all these years I have been pretending—that long ago I cut my soul in two like a bagel and offered my better half (the seeded half with lox and cream-cheese) to the Devil—and the lower half (with yesterday's kielbasa) to the elders.

I offered the Devil a continuing supply of suitably layered bagels. He could not get them for himself because of the hostile neighborhoods where the authentic ones are found: Although he is not welcome in those places, he knows them because he is ancient and has watched the long path from sand-infected unleavened bread to the succulence of smoked fish. I offered him the best: Nova-lox, succulent dripping sable, and large smoked white fish, which require separating the flesh from the bones. Then there were the half-sour pickles and the chopped-liver—oh my.

I was willing to act as the Devil's gastronomic emmissary (imagine all those charred sinner-meats that for eons he had to eat in Hell) but on the condition that he tell his anointed daughter (who holds Hell's creative portfolio) to give me the lead baritone role in that forthcoming didactic-demonic cantata: "The Concupiscence of Agony and Ecstasy." A New York run is planned—to open on All Saint's Day. The Devil's daughter is directly in charge (Hell is family oriented in all its dealings). During the first rehearsal, I did not do too well. They couldn't hear me In the back. I do acknowledge that I have a small voice and limited experience, but I also have a nice tone, a febrile imagination, good lung capacity, and ingratiating gestures. Believe me, Lady, if you give me a chance, I will grow into the part.

The Devil—as Devils do—has not yet answered my petition; the Devil's daughter will, I think, accept me. But while waiting let me remind her—as Faust should have (did you know they once were a number?) that this is not my first moment of free will. I can always sing in the subways! If she does not satisfy this last of my ambitions, then I will sing my songs—not in her murky labyrinths but hey, right by the local stop before the pearly gates. Better take advantage, then, Ms. Devil, of my present openness to offers. Look you and listen, you sweet and smoky lady: I can sing strife and contumely, "Cortigiani-Quel Razza Dannata," or I can sing rapprochement "In diesen heil'gen Hallen," or I can sing seduction, "Am Leuchtenden Sommer's

Morgen." Take your pick—but remember: My voice, even when I am dead, will eventually grow old.

Beginning

The accumulating wisdom that accompanies birth to death, dark to light to dark again, is exampled in the early transitions from the teat to the bottle and from the diaper to the potty. These changes include the intrusion of others into your self-sufficiency. Others like you and yet not like you come and go without your asking, taking away your waste, wiping off the stains, and offering something new in place of what they have just removed.

Increasingly, the changes pass from your world of touch and smell into the fuzzy shapes you see outside your window. With time, these magical extensions of yourself will turn into actual others who are not like the early ones—the ones who weaned and washed and made you into something special. It is now your time to be unlike those others. But however much you reach beyond who you are now, there are still more others—reminding you of the loss of intimacy and the start of separation over which you once had no control, and soon will lose again.

Nothing early on was as important as the teat and potty; but these memories of loss and gain did serve—in the middle years—to make sense of passing time and the growing number of people who show up and come into your life—each wanting something they will not tell. Even before that, the fear of leaving the dark and floating place where you were alone before time and change began, turned into the anxiety of being discharged into the world of others. You protested, with the vehemence of an embryo, against leaving that place where all was quiet and warm and nothing needed doing. Then soon, the first contractions began—and you kicked and screamed at the prospect of going further than you wanted to.

During my birth, I was pained by the interruptions to the constancy that I expected to have remain forever ("Forever" is not merely a notion of the fetus—it is a given—evoking no contraries in the womb). Then I was pushed to leaving without notice—although there were signs, had I been sharper, to alert me to that event: I was getting bigger, the womb was getting smaller, and the stillness was being broken by my kicks and the outside voices. I became irritated with all those changes, and began to anticipate—a major cognitive step—what my first move would be. But before I could think further, something began pulling on my head. Then there was a blinding light, and to my horror, I was hung by my feet and slapped on the ass until I took my first breath and cried. These were lessons quickly given but quite enduring in my new world—the first lesson was of pain, the second was the distinction between what I wanted and what actually happened.

But soon, I must admit, the assaults subsided and I was pleasured by soft voices and given the almighty teat to suck that made me full and soothed my gums. But this too did not last. The enveloping aura of eating stroking washing sleeping waking, gave way, slowly from the outside but too fast for me, into a separation of obligations: Leave the teat and eat the glop of carrots from a spoon; sit on the potty instead of shitting anywhere you like. The caressing hands-on was by now concentrated on mornings and evenings in the bath; and the kisses that came with drying and the smell of baby-powder signaled the start of a solitary night.

Then I came upon the wonderment of dreams. When—I don't remember—was my first dream and what was it about? My dreams continued from their daytime seeds into the nights where they replaced, with their thick presence and their absence of time, the fading daylight by the dark. But I could not control my dreams anymore than I could control my expulsion from the womb. Dreaming often made being alone at night more scary than the dark—but some dreams were full of sweet smells and the taste of milk and soft Zwieback. I never knew which would come when I began to drift. My daytime was full of directives about when we play and when we eat and when we potty, wash and go to sleep. But my dreams, once I was dreaming, did not tell me what would happen next, and none was ever tiresome, as were many of the daytime-stretches between my dreams. So I became two: one among obligations of the light, and the second the creature of my dreams.

I have heard that some people do not dream—or do not remember their dreams. I have always dreamt a lot, and my dreams often extend into my waking. I try to see these border-crossings as proper parts of living, and I accept the terrain on both sides of the gate with the wonderment due each—for reality is no less mysterious than dreaming. The uniqueness of waking, despite its claim to clarity, rectitude, and reality, becomes thin and sallow when not challenged by the other side of the show—by dreams and nightmares, visions and visitations.

Such distractions to the state of being thoroughly and confidently awake were admittedly a hindrance to my early life and did not help me in the battles with my peers. But these distractions became more important as I grew away from the neighborhood world of direct coping, and into the abstract demands of school. There I found reasons to step around the furniture of simple survival, and I looked upon the new demands with an eye sharpened by the bazaar of offerings that dreams provide. I wrote feverish term-papers in which I glorified my preference for fantasy over reality and listed ways I found to translate the day's events into suggestive myth. These were pretty much all I had to counter my earlier constriction in the depression days of Brooklyn. Writing turned out to be quite effective—for it separated me (I did not know this then) from the more practical thinking of my classmates. Drawing (which I also did a lot of) was an even more potent separator. My classmates took on the privilege of making remarks—judgments too—about my drawings. Beneath contempt, they were. No-one but some teacher read my writings.

In college I took a class in modern poetry which was taught by a beautiful woman whose passion was for Teilhard de Chardin—God and Modernism together. I wrote papers that she praised, until the one she should have praised most—for it coupled God and she and me, after all, the Trinity. She did not respond to it—she gave it a "B" and when I asked her why, she just said it was not very good. I had the urge, the first outside my dreams, to kiss her. I now know she knew and feared—outside of poetry—that amalgam of penitence and lust. She, of course, gave me an "A" for the course, and would greet me by my last name when we passed each other in the hall.

Eventually, the larger academy became impressed with my disjoint forays into learning, and this started a long and pleasantly enduring time in my life, like the early period of my gestation. When it came to an end, I could only recognize two ways to go—the one towards a quiet circumspection filled with proper and peripheral notions, the other a re-birth that would echo my first birth, and propel me toward bright lights, noise, and the latest amalgam of pain and pleasure. But I did not have the courage for an "Either-Or"—no one had really told me that in life I had to choose—so I hopped and skipped between the two—jumping aside at the second hint of non-refundable pain or the third sign of a too demanding pleasure. This discretion probably saved me from dangerous reaching—also from untold bliss.

Pleasure is seldom writ out-loud. This is perhaps because descriptions of loud pleasures entice us away from the more delicate ways of reading and listening, and lead us into rough daytime of instant gratification. When we are still nimble, literary pursuits seldom satisfy our appetite for the coarser stuff. But if we persist in finding our dream-desires, we must challenge our appetites—smite them at their conventional roots—give up on direct eating and increase the tasting. The youngest and lightest of us run around a lot; I did not persist in running—good thing, for I doubtless would have fallen off a cliff, never to be heard from again. Oh, I am reasonably well coordinated, and I did try some approaches to a moderate mountain—but then I slipped. So I began writing about how it is for those who are not slip-prone—or for those who persist slipping.

Actually, it was better that I slipped so soon—tumbling my then skinny ass across the slanted rocks—rather than tripping later when on a less forgiving slope.

I remember passing the bony beautiful mini-bikini Euro-lady lying on a chaise, perched on a snow patch, fronting the last edifice of mountain-lodge civilization before the split between the peaks of the Monch and the Jungfrau. Just beyond her hauteur lay a considerable test for novices—not too dangerous—but enough of sheer drops on both sides for me to recognize the value of my guide. He told me to go ahead of him so that he could rope me in if I slipped, but if I were to follow him, he said, my slip would pull us both down. So we slippers do survive to spin our tales of conquest on the high mountains such as might entice others away from their reading—like that lovely skinny unfriendly lady who should have looked up from her preoccupation with her tan, and moistened the rarified air of my mountain triumphs.

She was gone when we returned. Too bad. I had that desireable aura of grime and pallor that befits the mountain adventurer. But by our return at twilight time, she was warm and liquored, freshly bathed and perfumed in her filmy post-bikini wrap, sitting demurely in the lounge (which is known for its world's-best mountain view) where she would meet with her rich and properly sedentary older lover. For a while, fat fingers would stroke tan flesh; then they would gossip about the summer affairs of seasonal friends.

But even if you had replaced old fatso with your skinny agile fingers, you and the lovely lady would both have been unhappy. She would have criticized your French and severely dented your self-love, thereby pushing you off the edge—a bit too soon.

Better instead to remember the tableau-vivant of a certain cocktail party (the one celebrating good art) that continued on well beyond the last potato-chip into the scenario of mingled smells and the shrieks and giggles of late-night frolics. All this came to its natural end when you and I, saturated with each other's sweat (remember, we selected each other around ten that night) stumbled into the early morning light to watch the downtown brokers whiz past, riding high on Harleys. We thought then that we are fortunate—without the drag of money or motorcycles—to have experienced (just a while ago) the meeting of pain and pleasure well-jelled with not a few smelly and yelly orgasms. Pleasure (we agreed) as we stood naked in the morning, is about taming the loneliness of pain. Each, when seen in the light, is the other in disguise.

Middle

Here-and-Now (where-ever these may be in time and place) is the crucial moment for directness. I am, at an advancing age, still healthy and still firming the alliance between what I did not do and what I have yet to do—the stuff you are reading now.

Pleasure can be expansive out of doors. During our vacations we slip and slide across the rocks and think the waves will not reach us—only their spray wetting down our suntan oil. Below, the sea batters, the rocks shake, and yesterday's storms illuminate the changes. Optimism is a necessary condition for pleasure—the waves cannot reach us here—we paid enough to get on high—and the rocks below us will endure. But, as with the end of summer, the folderol of fun does not last beyond its season. Late pleasure, like tired rocks that must face winter, breaks down and in sober moments comes to see its end. During the first advent of its beginning, pleasure, like the new-born babe, does not worry that it may not continue. Later, faced with the time when seasons change, and nerve-endings are benumbed, pleasure ends its engorging spin and accepts replacement—not by the scolding and lamentations echoing over Puritan Ridge—but by ordinary things: the level mix of indoor comfort, warm fireplaces, and home-cooked meals.

"Ordinary" does not of course, say it all. Despite the winter's aridity, we long-time patrons of the hot and wet hope that the summer gift of sunburn and the soothing plays of lotion and seduction are not completely gone. These may well fade with season's end into scaly skin and back-biting—the accoutrements of winter. But the temptations of summer—if one is committed to their juices and their sounds—can be made to continue through the cold of winter. But this requires work: Phone-calls must be made at discrete times in the short-dayed months—preferably between ten and three on weekdays; then there are the packages with toys of love to be stashed at arranged places for occasional use.

The unencumbered meadows of our summer thing is where the fear of ending is most apparent. While having been an end before, ending is still surprised by the sudden gust of cold that is the imprimatur for puckering-tight, cleaning-up and packing-out. Unlike midsummer pleasures, with their insistent and unheeding thrusts, the winter life—the one where we retreat to radiator heat and slushy streets—seeks little action beyond conversation about the unwelcome changes that had taken place last summer on our once pristine beach. But unlike the atonal arias we sound into the wind when early summer pulls, we speak quietly in winter, so as to not disturb the neighbors who live behind the porous walls separating us from their next-door condo.

We could, we say, go back to the sea right now—but disappointment always clouds the return to your summer home in winter. It has to do with the loss of ambition and the acknowledged weariness of middle age. What is there left of ambition now that will not be drowned by the sound of winter waves? And those other desires—money, fame and love—just mark my words, will be diluted faster there than here. But staying here can be worse. No more I say, to these city winter parties—every one looks so fat and old. I hate their faces—so smug and saggy, and I hate their places—so full of the tasteful objects I despise—a compendium of grandma's leavings and the sale at "New-Age Antiques." And I won't either make a party for them—those perfumed shits—who would eat my good food, drink my passable wine, bad-mouth my virtues, and ignore my vices.

Feeling noble and much misunderstood, I can indulge my winter tantrums when I stomp the snowy streets and expound my views into the teeth of the snowy gale. I reach high-C as I proclaim my independence from those too-sweet people and their tacky stuff, and I make wild paw-prints in the snow. There is nothing better than a good stomping in fresh snow to purify the mind.

How did I get back here—back by the sea? But I see you here beside me. Are you Heidi, who despite your now-large thighs, are still the ice-queen of legend and the Hamptons? Were you hiding your love of cold when we first cuddled naked by the fireplace—you did sweat a lot, as I remember—and were you not planning all that time to take my hot-rod and me to the frozen shores, for a midnight plunge?

Yes, she said—all that. A cooling-off period generates a novel heat; and the sea has a thermostat that you can regulate according to your need. Smart woman, Heidi.

So on a cold and windy day, we go back to the house we built that looks diagonally onto the sea. We do not come there year round, because we are busy oh so busy with our futures in the city. Remember, though, how it is in early summer when ambition begins to wilt by afternoon? You endure the outbound traffic and then turn off onto the country road leading to the graveled path which ends at the small house with large windows and two decks. There are no signs of break-ins; the trees and bushes have gone through their first leafing and the flowers are beginning to show. But there still is the nippy wind propelling us inside, there to make the sounds that wake the furniture and alert the others—pots and pans and garden tools—that we are back, ready to use them for our ends. The house itself is girded by stone barriers and shingle walls, meant to protect us from the larger waves that are expected most winters. In the summer, all this can look quite overbuilt—like a great castle with its retaining walls. But in winter, what with recent shifts in climate and theory, the house begins to look flimsy—like a child's attempt to hold back the sea.

There are storms every winter now. The winds push the cold waves against our already crumbling rocks. These are the waves (if we go beyond the signposts we erected on the cliff) that could wash over all of us and take us to that cold roiling place of sand, stones, and seaweed. So we warn the children to stay away—and please watch the dog too.

But what is there to do in winter at the sea-shore when the sun sets early, the wind howls, the phone doesn't ring, and it's only six o'clock? Well, there is always the option of leaving that lonely place (I never really liked it, you know) and returning to the city with the spouse who is now more flabby and prickly than when she played "summer seductress on the sands." But that was a while ago.

If we stay however, we could do more than listen to the cold waves, we could walk to them barefooted in the snow—or we could watch some pornography after supper. Don't be dismayed. "Porn," you know, has been called a "minor art" by people who should know. Some even say its redeeming virtue lies in exploring the fragile barrier between character and performer. That's interesting! I must admit that I never know whether you're actually coming when you make those sounds. On my side I do show some sticky evidence—but still I'm never sure that I know exactly who I'm with or where we are. Porn tells us something about ways of having sex, but little about what is happening behind the sounds those contorted faces make.

Unlike the demands we make of "legitimate" theatre, where good actors play muliple characters, we want the shows of sex to be real. But this harks back to "method-acting," a theory which instructs the performer to "live" the role being played. It turns out that, as a method, "living the role" had the actors playing themselves and not their character—freedom became monotony. But is it not this identity between self and character just what should occur in pornography? To the contrary, one might think that porn-actors need only the ability to show—convincingly—that whatever else they feel, they have climaxed before film's end. "Only showing" should be good enough for casual viewing.

However, this genre is distinct from traditional theatre in that its connoisseurs want the performers to be true to life—to the experience as well as to the act of what they're doing. This is an argument for porn's uniqueness as an art: If you will, pay close attention to every sound and movement—see the moment(s) when the performers have transgressed the border between fiction and fact. Pay particular attention to inadvertent breaks and flubs in the routine—changes of position while in the so-called center of things; glances to the side to see what the director wants; a sudden blank stare or a loosening or tightening of facial muscles, a spasmodic jerking of limbs. These are all signs of simple needs (perhaps the need to pee or the need to come) which deny the fictional hoopla called for by the script.

For the sake of the argument, it is fortunate that much pornography is low budget and cannot afford the niceties of expert illusion or multiple takes, and so suffers a lack of control—permitting the intrusion of the actuality which breaks through the scripted play. This porous divide between represented sex and lived sex reveals what is most valuable in pornography: The genre has its claim to art by showing the aesthetic impasse—the unstable distinction between art and life—the unique moments when priapic fiction becomes real by the actor's actual need to be orgasmic.

The performance of sex always has a viewer, whether it is the distanced gaze of others in a theatre, or the divided attention one pays in one's own bed. We are all performers in our sexual places—whether in the private mirrors of our mind or into the windows of an audience. We watch ourselves as we watch others, and they watch us while watching themselves watch us . . .

Parenthetically, I appreciate pornography for its presenting images of large breasts and wide hips. I have little interest in fashion, but I do abhor its anorexic look—a subtle form of coercion. In contrast, I think amplitude to be particularly beautiful. I enjoy looking at works from the great masters of the old salacious days—Titian, Rubens, Courbet, Renoir, also Schiele and Lautrec—how seamlessly they could join the subjects of sexuality with the requirements of great art. The images of fashion are hopeless in this regard—but pornography could learn much from the older traditions: Amplitude is inimical to repetitiveness and posturing, but it can flourish with age, austerity and simple elegance.

I find the older ample ones to be well-creased—good to draw—the way I draw.

Beneath your robes, my love, you show the small ravages, the creeping blemishes, the sags and wattles of age. You are not smooth like the ancient marble that once they used for stroking, but you show what we must find when we really want to look.

In the images of erotic art ('Pornography' is actually a too-hostile word), I look for the moving moments of confession—the intonations and exculpatory cries of excess and indulgence. They are the music and lyrics of sex: "Believe that you can know my soul the ways I show it—I will believe the same with yours." There is pain of course in these beliefs—the pain of hubris—which is found in the fulfillment of present ambition when it ranges too far from later truth.

End

Time passes and I go from place to place. But I hope it comes to pass that the passing of time gives way to the places I am at.

Places change as the head turns. Some appear and others leave as I walk around. But it is the place that matters—not the getting there. We all want to be in a good place—a peaceful place, an interesting place, a warmer or a colder place, a place of one's own, a place where we all can be together. There also are the memory places—although they always move too quickly past—that I conjure out of glances, dreams and recollections.

Time is the collaborator with ambition and shows the way to go before one grows old. Time once marked the stretch from farmhouse to townhouse—it is a goad to achieving what one wants (needs) to fill a life.

Towards the end, however, time gets in the way of places that will open when you pay attention— places where you might (should) want to stay.

Death needs time as tomatoes need salt.

The cocoon of places one has built, encourages death's irrelevance.